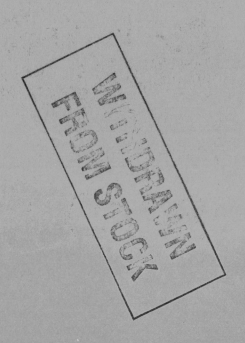

# THE METALS OF LIFE

*The Solution Chemistry of Metal Ions in
Biological Systems*

# THE METALS OF LIFE

## The Solution Chemistry of Metal Ions in Biological Systems

DAVID R. WILLIAMS

*Lecturer in Chemistry*
*University of St. Andrews*

VAN NOSTRAND REINHOLD COMPANY
LONDON

NEW YORK    CINCINNATI    TORONTO    MELBOURNE

VAN NOSTRAND REINHOLD COMPANY
Windsor House, 46 Victoria Street, London, S.W.1

INTERNATIONAL OFFICES
New York   Cincinnati   Toronto   Melbourne

Library of Congress Catalog Card No. 77-125957
SBN 442 09499 X
First published 1971

PRINTED IN GREAT BRITAIN BY
THE CAMELOT PRESS LTD., LONDON AND SOUTHAMPTON

# PREFACE

IF ONE meddles with the normal processes of life, there is the possibility that it might result in causing death. Nevertheless, people nowadays are caring more and more about people and it is reasonable for chemists to take an interest in the chemistry of people and to see if they can improve one's lot. The biochemist's role is in understanding, not tampering with, normal living processes so that the wayward reactions of the sick can be reinstated to the straight and narrow. With the increased emphasis on welfare and medication, life is no longer a matter of survival of the fittest but rather a higher emphasis is placed on sustaining the unfit. This means there are becoming more and more sick people in the World and so the search for prophylactics rather than palliatives must be increased. This book endeavours to show how inorganic chemists can share in the enthusiasm and excitement inherent in the search for the relationships between chemical structure and biological activity.

The urge to take drugs is a notable characteristic that distinguishes man from animals. This desire has been inbred over the several hundred years during which witchcraft has been developed into the very advanced fields of medical sciences. Medicine today is in such a high state of perfection that improvements are difficult to come by. One attitude adopted to research for these improvements has been the investigation of the orderliness of nature and the implication of ordered attacks upon disease. One such basis for the order of nature is that founded upon the role of metal ions in living systems. We hope to introduce students to this order.

It is highly probable that discoveries of the future will be made amongst the interdisciplinary areas of science. Nature has never divided herself into the sharply defined areas of chemistry, biochemistry, physics, etc., but rather she has used all the advantages of each branch to design a unique, superbly produced, product. This book is written to bridge the gaps between inorganic chemistry, biochemistry and medicine and to introduce students of one of these fields to the other two. Hence, one-third of the book will seem childishly simple to each student, the other two-thirds, we hope, will appear challenging. These three branches are being viewed through the metal ion approach and, in this way, it is purported that interdisciplinary barriers can be removed.

How can mere chemists contribute to this process? On the one hand, they can investigate the basic principles behind nature; for example, it was at one time thought that biological systems disobeyed the second law of thermodynamics, whereas, after extensive researching, it is now known that, although living systems are always fighting against the second law—the energy for this fight coming from a series of coupled reactions—life still obeys this law. On the other hand, it is well known that nature is not accumulative. A compound of the two species A and B does not only have the composite properties of both A and B but also it has specific properties and features of its own. Hence future research must be directed towards understanding (and this includes being able to predict) properties of composite systems with respect to their sub-units. In this context metals are unique in that many sub-units are compounded into larger units only by and through metal ions in metalloenzymes. Hence, metal protein complexes emerge as a central problem of biochemistry. How may they be characterized? What is their stability? How do they function in biological systems? We must bear these questions in mind as we work with metal ion solutions because only with our eventual goal in perpetual sight can we hope to find the pattern and the pathways that lead to it.

Although the majority of elements in the periodic table are metals, nature has chosen to use mainly non-metals as her building blocks. However, some metal ion biological chemistry has been known for a long time. For many decades it has been held that all the metal ion *in vivo* chemistry was thought to be tied up with the hydrations of the bulk metals sodium, potassium, magnesium and calcium which accounted for 99 % of all the metals in our bodies. Their biochemistries were basically those of hydrated ions satisfying necessary osmotic pressure requirements and giving strength to our skeletons. More recently interest in the trace metals—the transition elements—is gaining ground. Because their *in vivo* chemistries are newer they will be given priority in this book.

It has always been hoped that chemists use their discoveries wisely and now that *homo sapiens* has so much freedom and so many opportunities to control his own destiny it is also desirable that the non-chemist should know enough chemistry to anticipate the outcome of using new, or combinations of, chemicals or drugs. History has never witnessed such a time when it has been more important for laymen to understand some chemistry and for chemists to understand the mechanisms of *in vivo* processes.

In writing a book that is designed to point students, not only of chemistry, towards the bridges to other hunting grounds I have had to use my judgement about the content, organization and relative depth of coverage. What is the use of science if we cannot communicate the salient ideas even if this means, in the first place, setting aside the mass of facts and references that lead to these ideas? For clarity, only summary references have been chosen, original results being avoided. This is not, please *not*, to decry the thousands of more factual references and the millions of laboratory hours given by devoted scientists to determine the facts upon which the principles in this book are firmly founded. Words cannot express our debt to the unmentioned. In order not to place a toll on these interdisciplinary bridges a glossary of technical terms is placed at the end of the book.

*St. Andrews, March, 1970* D.R.W.

# CONTENTS

*Chapter 1*

# INTRODUCTION

---

1.1 *The superlatives of* in vivo *chemistry*

IF ONE is medically motivated to apply chemical principles to bio-chemical knowledge, there is a chance of making scientific progress. However, there must surely be no one who claims to know all the chemical principles, all the biochemical reactions and all the medical aspects of the human body. Even so, the author endeavours to introduce the implications of these three branches of the science of the 100,000 processes (each being a compendium of many chemical reactions) that occur in the human body.

Ordinary life is chemically extraordinary in that ordinary rules cannot explain the very high bond strengths, whose magnitudes are so often associated with reactions at extremes of temperature and pressure, occurring at 37°C and 1 atm pressure. Abnormal life (e.g. anaemia, cancer or rickets) is even more extraordinary. Scientifically, the aims must be (a) to try to understand the thousands of reactions met in ordinary life, (b) to investigate how these can deviate from normal and (c) to attempt to discover methods that can reinstate the erring reactions in (b) back to the straight and narrow. These three are mutual; the deeper our understanding in (a) and our investigations into (b), the greater the probability of method (c) being a cure rather than a palliative crutch.

Unfortunately, nature is much more scientifically advanced than is research. How often has one attempted to conserve energy—inside an adiabatic calorimeter, insulating a house, or by trying to produce an effective accumulator or fuel cell? And yet, the human body has the means of doing this, and furthermore, it can regenerate this energy, when required, in 100% efficiency. The author refers, of course, to the sophisticated ADP–ATP system.

To control such systems the human body contains the most effective and specialized catalysts known to science. These are called enzymes and, for its own reactions, each enzyme is an ultra-efficient catalyst whereas for any other similar reaction its efficiency may be zero. Further, they have the attribute that they can completely regenerate themselves. These enzymes contain very important cavities called 'active sites' and the power of the enzyme depends upon the three-dimensional geometry of the chemicals lining these sites and upon the induced electronic environment around them. Although metals are by no means always present within these sites, enzymology does seem to be more advanced in the metalloenzyme field as instrumentally they are easier to study because the metal ion provides a convenient electronic heavy atom marker. Enzyme reactions involving metals are of two broad types—those in which the metal ion comes and goes from the active site and those in which it is strongly bound therein and does not leave. As might be expected there is a large overlapping hinterland between these extremes.

Because of the exceptionally complicated environment around the metal ions in metalloenzymes one must expect the unusual or exceptional when investigating such systems. Clearly these studies are best done by combining the skills of highly trained specialists (scathingly defined as 'ignorant in every subject save one!') from many disciplines, each one broadening his research profile and possibly still maintaining a depth of knowledge as each expert synergistically teaches, helps and encourages his colleagues.

## 1.2 Why inorganic solution chemistry?

A decade or so ago it was the widely held view that enzymes were all organic compounds that were crystalline, colourless solids. However, some were yellow, blue, green or greenish-brown. Most of them were soluble in water or dilute salt solutions. It is now known that these transition metal-like colours often arose from the transition metals present in the enzyme and that the solubility properties are metal ion dependent also. Finally, some metal ions cannot be removed without destroying the structure of the enzyme.

The following reactions are just a few that have definitely been shown to be transition metal dependent [1].

1. $O_2 \rightarrow H_2O$, $N_2 \rightarrow NH_3$, catalysed by $Fe(II) \rightarrow Fe(III)$, $Cu(I) \rightarrow Cu(II)$.
2. Dehydrogenation, e.g. ascorbate → dehydroascorbate, catalysed by $Cu(II) \rightarrow Cu(I)$, $Mo(VI) \rightarrow Mo(V)$, $Fe(III) \rightarrow Fe(II)$.
3. $H_2O \rightarrow O_2$, catalysed by valence state changes of manganese.
4. catalysed by $Cu(II) \rightarrow Cu(I)$
5. Ribose → deoxyribose catalysed by $Co(I) \rightarrow Co(III)$.

Our present knowledge of these inorganic energy generating catalysts is so elementary that synthetic copies usually cannot be made. However, it might be noted that nature has been thrifty in her designs in that many enzymes having similar roles have a common metal ion held to the active sites by common donor atoms. An early classification along these lines is shown in Table 1.1. Similarly we find that there is further thrift in that the peptide

TABLE 1.1. General role of metal ions in biological processes

|  | Na, K | Mg, Ca, (Mn) | Zn, Cd(Co) | Cu, Fe, Mo, (Mn) |
|---|---|---|---|---|
| Type of complex bonds | weak | moderate | strong | strong |
| Biological functions | charge transfer, nerves | trigger reactions, hydrolysis, phosphate transfer | hydrolysis, pH control | oxidation and reduction reactions |
| Ligand atom preferred | O | O | N and S | N and S |

chains forming the site for the metal ion are sometimes made up of sequences of amino acid residues that are common to another enzyme.

## 1.3 *Reaction vessel and ingredients*

Our blood streams are the sites of intense metabolic productivity and the composition is continually changing and being renewed and adjusted as constituents are formed or destroyed (their average life time is only hours or days). Nevertheless, chemists require to know the dimensions of the vessel and the approximate contents. Such information is given in Table 1.2.

TABLE 1.2. Concentrations of some of the constituents of human blood. Many of the figures have been calculated from those listed in *Biochemists' Handbook*, Ed. C. Long (E & F. N. Spon Ltd, London, 1961), which can also be used as a source of (a) the errors in the values listed below and (b) concentrations of the constituents found in other bodily fluids and tissues.

Blood = Red cells (called corpuscles) + white cells + platelets + plasma
(92% $H_2O$)                                  (these float in plasma)

Plasma = fibrinogen + serum

| | | | |
|---|---|---|---|
| Sodium | 85·2 mM | in (whole) blood | adult |
| | 141·3 mM | in plasma | |
| | 140·1 mM | in serum | |
| | 20·9 mM | in cells | |
| | 137·8 mM | in serum | newborn |
| | 20·9 mM | in cells | |
| | 126·1 mM | in serum | foetus |
| | 36·1 mM | in cells | |
| Potassium | 44·5 mM | in blood | adult |
| | 4·11 mM | in plasma | |
| | 5·06 mM | in serum | |
| | 94·9 mM | in cells | |
| | 5·4 mM | in serum | newborn |
| | 151·7 mM | in cells | |
| | 10·2 mM | in serum | foetus |
| | 158·6 mM | in cells | |
| Magnesium | 1·57 mM | in blood | adult |
| | 1·13 mM | in plasma | |
| | 0·87 mM | in serum | |
| | 2·72 mM | in cells | |
| Calcium | 2·42 mM | in serum (adult) | |
| | 1·85 mM | in serum (5-day-old infant) | |

| | | |
|---|---|---|
| Manganese | 2·18 µM | in blood ⎫ |
| | 0·73 µM | in plasma ⎬ adult |
| | 1·46 µM | in cells ⎭ |
| Iron | 8·59 mM | in blood (adult) |
| | 23·3 µM [59·09 µM] | in serum [T.I.B.C.], see p. 16 (adult) |
| | 17·5 µM [84·15 µM] | in serum (mother at parturition) |
| | 30·9 µM [46·37 µM] | in serum (infant cord blood) |
| | 18·6 mM | in cell (adult) |
| | 17·7 mM | in cell (newborn) |
| | 17·0 mM | in cell (foetus) |
| Cobalt | 0·71 µM | in blood (adult) |
| Copper | 14·8 µM | in blood ⎫ |
| | 18·3 µM | in plasma ⎪ |
| | 18·1 µM [16·6 µM] | in serum (male), [female] ⎬ adult |
| | 11·9 µM | in cells ⎪ |
| | 42·3 µM | in serum late in pregnancy ⎭ |
| Zinc | 138·4 µM | in blood ⎫ |
| | 47·2 µM | in plasma ⎬ adult |
| | 226·5 µM | in serum ⎭ |
| Molybdenum | — | no cations in blood |
| Bicarbonate | 28·8 mM | in blood (adult) |
| Chloride | 80·9 mM | in blood ⎫ |
| | 100·9 mM | in plasma ⎬ adult |
| | 105·7 mM | in serum ⎭ |
| | 108·6 mM | in serum (newborn) |
| | 106·6 mM | in serum (foetus) |
| Serum | 70 g l.$^{-1}$ | adult |
| | 62 g l.$^{-1}$ | at birth |
| Free amino acids | 4·14 mM | Estimated from total free amino-acid nitrogen. For a complete listing see *Biochemists' Handbook*. |
| Blood volume | 5–6 l. | 1 pt per stone of weight |
| pH | 7·35 – 7·42 | |
| Temperature | 37–38° | |
| No. of red blood cells | $5 \times 10^{12}$ l.$^{-1}$ | |
| Aver. life of a red cell | 120 days | |
| No. produced per second | $2·4 \times 10^6$ | |

Although there are means of regenerating enzymes over and over again, some metals do get excreted and so a means of replenishing them has been provided (called ingestion). The steady state concentrations in Table 1.2 have all arisen by the relationship:

Total ingested → Total in blood          → Free concentrations in
(diet dependent) (filtration controlled    blood (dependent upon
                 in stomach wall,          the equilibrium constants
                 kidneys, etc.)            of metal ion ligand com-
                                           plexes)

These three quantities can all be medically controlled:

(a) Total ingested—can be regulated by chemotherapy or dietetics.
(b) Total in blood—the filtration stage can be bypassed by an intravenous injection.
(c) Free in blood—can be increased as in (b) or reduced by adding sequestering reagents to mop up the excess free metal ions.

In Chapter 2 these metal ions are further discussed, and in Chapter 3 the range of ligands involved. However, we ought not to get an oversimplified view of chemotherapy, or aim (c) on p. 1 as, although 90% of the medicines prescribed by doctors today were non-existent fifteen years ago, the thousands of new drugs a year approach is not quite correct: the average number of new drugs accepted by the health authorities each year is about forty. Approximately 10,000 are tested for every one accepted. Clearly here we have many potential new ligands to complex with our metal ions and every new possible drug must be carefully screened. Pharaoh Cheops built the largest pyramid in Egypt by employing 100,000 slaves (average working life expectancy of 2 years) who laboured continuously for thirty years. In order to screen these new drugs, slave replacements must be invented, e.g. a countercurrent distribution apparatus can perform the work of 250 men; computers can be used to replace many men's brains. Nevertheless, even using these new techniques, the work is still of gargantuan magnitude, e.g. just one application for approval of a new drug consisted of twenty-seven volumes of laboratory reports involving the work on 4500 patients by 200 investigators [2].

2000 chemical reactions can occur within one cell of a human body. Scientists performing screening tests must ensure that none of these reactions are unintentionally upset by administering a new drug. Thanks to their efforts, tuberculosis, pneumonia, influenza and tropical diseases need no longer prove fatal.

## 1.4 *Our limitations*

Our bodies contain most of the elements of the periodic table (many in minute or trace quantities), e.g. mercury from dental work and newspapers, metal ions from cooking pots, assorted pesticides and preservatives from surfaces of unwashed fruit; but to keep this review to manageable size, only the metal ions actually required for life are discussed. However, the strength of protein–metal ion bonds are such that any metallic cation that finds its way into our bodies will spend a large part of its time bound to amino acids and proteins. These impurity or poisoning reactions would come under (b) in the second paragraph on p. 1.

As already mentioned, proteins live in a state of permanent flux in which they are continually being broken down and re-formed. They build, transmit and destroy and are themselves built, transmitted and destroyed. Compared with other parts of the body, the blood is a relatively quiescent area and so most solution chemistry research has been concerned with the blood stream [3].

Microscopically speaking, an enzyme is known to form very strong bonds to its metal ion; enthalpy measurements enable us to measure the strengths of these bonds. Macroscopically, our bodies are mechanically strong and the basis of strength is an organized arrangement of materials. It is this organized arrangement that we must look for in metal ion bodily ligand reactions. The reader will find gaps in the correlations attempted here. If these gaps promote further research or interest then the author's efforts have been worth while.

## REFERENCES

[1] WILLIAMS, R. J. P., *R.I.C. Rev.* 13 (1968).
[2] BOGUE, J. Y., *J. chem. Educ.*, **46**, 468 (1969).
[3] TAYLOR, R. J., *The Chemistry of Proteins* (Unilever Educ. Booklet, 1969).

*Chapter 2*

# METAL IONS *IN VIVO*

---

A LIST of the metal ions found in our bodies is given in Table 1.2 where also will be found details of the reaction vessel. Built into this vessel we have a system for purifying the circulating chemicals and for regenerating them with oxygen gas. Details of these cleansing processes are given in Table 2.1. One might note the high efficiency of these processes in allowing chemicals to be used over and over again.

TABLE 2.1.  Details of the human body's cleansing and regeneration processes. Taken from reference [1]

---

LUNGS

    Surface area of alveoli $= 100$ m$^2$
    Fresh air entering alveoli $= 5\cdot25$ l. min$^{-1}$
    Blood flow through lung capillaries and other bodily tissues $= 5$ l. min$^{-1}$ at rest increasing to 30 l. min$^{-1}$ accompanying severe muscular activity.

KIDNEYS

| | *Filtered* (*day*$^{-1}$) | *Excreted* (*day*$^{-1}$) | $\therefore$ % *conserved* |
|---|---|---|---|
| Water | 170 l. | 1·5 l. | 99·2 |
| Sodium | 24 g. ion | 0·1 g. ion | 99·5 |
| Chloride | 18 g. ion | 0·1 g. ion | 99·4 |
| Bicarbonate | 5 g. ion | 0·002 g. ion | 99·9 |
| Glucose | 0·8 moles | 0 | 100·0 |
| Amino acids | 0·5 moles | negligible | 99·9 |

---

The metals are found both in solid and in solution states in various parts of the body and a cross-section of these metals exists as free ions and as their complexes in the blood stream.

It may be observed that some metals that are relatively abundant on the Earth's surface are not found *in vivo*, e.g. chromium and nickel. The suggestion has been made that these metals are so strongly held in minerals that they are unable to enter plants and then our bodies. If impurities do manage to enter the blood stream they usually become attached either to phosphate groupings of the nucleic acids or to proteins.

Each of the metal ions will be considered in turn and usual types and configurations of bonds found *in vitro* will be noted. In Chapter 4 and subsequent chapters it will be seen how much these characteristics carry through to, and how much they are altered in, *in vivo* reactions.

## 2.1 Sodium and potassium

As might be expected from their position in group I of the periodic table they are mainly ionic in character and form colourless ionic solutions and salts as solids. They do have some covalent character, sodium being usually more covalent than potassium, e.g. $Na(NH_3)_4^+$ is tetrahedral having Na—N bonds equivalent in strength to Zn—N in $Zn(NH_3)_4^{2+}$. Even in the sodium tetraammine ion the bonds have a degree of ion dipole forces rather than pure covalency. Similarly, the sodium and potassium ions are strongly solvated in water. By analogy the $Na^+$ has $4H_2O$ in its primary hydration shell and the $K^+$ has 4 or 6 (tetrahedral or octahedral, respectively). When both the primary and secondary hydration shells are considered together the $Na_{aq}^+$ is, surprisingly, larger than the higher atomic weight $K_{aq}^+$ ion. The hydrated radii of $K_{aq}^+$ is 0·232 nm (corresponding to about $10H_2O$) and $Na_{aq}^+$ is 0·276 nm (about 16·6 $H_2O$).

The alkali metals derive their name from the Arabic for plant ashes (*al-qily*) as ashes obtained by burning desert plants contain considerable amounts of alkalis and so we might expect (correctly) that our bodies obtain their supplies from plants that we eat, e.g. fruit and vegetables. Although the two elements have approximately equal abundances on the Earth's crust, plants have about ten times as much potassium as sodium therein. This has two outcomes: first, because of the rapid uptake of potassium from soils there is a need for potassium salts as plant fertilizers, and secondly, we can see why our diets have to be balanced by added sodium chloride.

The ions are found widely distributed throughout the body, sodium being the main cation in the fluids outside cells and the potassium being found mainly inside cells. The shock that occurs after severe burning is observed because potassium ions are lost from within cells. Both ions have the role of keeping the osmotic pressures on either side of the cell wall constant and they also maintain the sensitivity of the nerves and control the muscles, e.g. sodium ions depress the activity of muscle enzymes and are required for muscle contractions; potassium ions permit the heart muscles to relax between beats. Sodium chloride is the source of hydrochloric acid for gastric juices and sodium bicarbonate is a buffer in maintaining the acid/base balance of body fluids and in the transport of carbon dioxide. Sodium, as sodium chloride, is wasted in large quantities through perspiration. Both alkali metal ions are lost in the kidneys and then in the urine as the $Na^+$, $K^+$ or $NH_4^+$ salts of phosphoric, sulphuric and organic acids, e.g. uric or lactic acid

The sodium salt of uric acid is rather insoluble and when deposited in the cartilage produces gout.

It is surprising that there is a high concentration of potassium in blood cells and tissue and a low concentration of sodium ions whereas the fluids bathing these cells and tissues are low in $K^+$ and rich in $Na^+$, especially since $K^+$ and $Na^+$ can diffuse through cell walls [1]. The natural tendency would be for the sodium to diffuse inwards and the potassium outwards (see Fig. 2.1(a)). This does, in fact, occur but their movements are counteracted by an active chemical reinstatement mechanism that may be explained either by the presence of 'super' water inside the cell or by a process that takes place in the cell membrane and is called a 'sodium, potassium ion pump' (Fig. 2.1(b)). This plucks a metal ion out of a region of low concentration and expels it into one of higher concentration. Because of the imbalance of metal ion concentrations on either side of the permeable membrane a potential difference develops and so the net outcome is that the ion pump mechanism has to expel the sodium ion against a concentration gradient and an electrical potential. If we switch off the

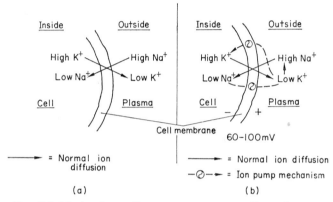

FIG. 2.1. Mechanisms of ion transport across cell membranes.

ion pump by freezing its enzymes (e.g. to 0° C) normal Donnan membrane equilibrium prevails, Fig. 2.1(a), and the concentration of ions become equalized on both sides of the membrane. This happens in low temperature theatre operations and in blood transfusion vessels stored in the cold. On rewarming the blood reverts to its normal imbalance of ions, Fig. 2.1(b).

## 2.2 *Magnesium and calcium*

These two ions of group II of the periodic table are isoelectronic with the two ions just discussed from group I. However, apart from the hydrated radius of the $Mg^{2+}$ ion being *larger* than that of the $Ca^{2+}$, there is no other large resemblance between the two groups. Calcium and magnesium ions weigh more and are only two-thirds the diameter of their group I counterparts. This increased nuclear charge density makes them much less polarizable. In aqueous solution the ions are octahedrally surrounded by a sphere of $6H_2O$, e.g. $Mg(H_2O)_6^{2+}$. $Mg^+$ is a very short lived ion and so there is negligible redox chemistry for the $Mg^+/Mg^{2+}$ couple.

Simple ligands form covalent complexes such as $MgBr_2$ and $MgI_2$ with magnesium and $CaCl_2$ is fairly covalent also. When we turn to the more complicated ligands those having oxygen donor atoms are preferred. These are found in phosphates or in the ethylenediaminetetraacetate ion (Fig. 2.2).

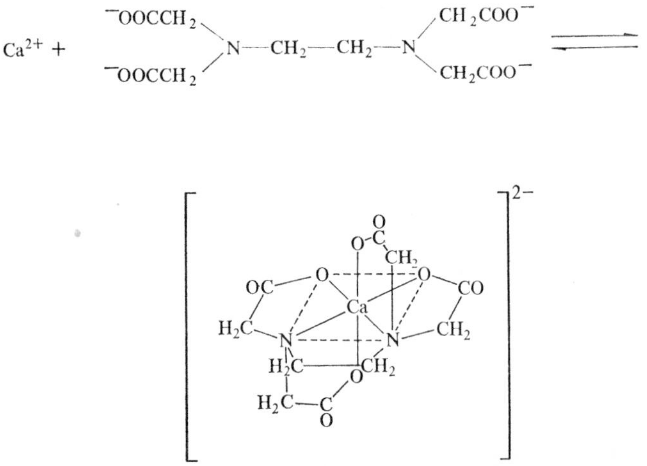

FIG. 2.2. Calcium ions forming an octahedral complex with the ethylenediaminetetraacetate ion.

It is easy to satisfy our metabolic requirements for magnesium and calcium as they are among the eight most abundant elements on the earth's crust. Magnesium is found covalently bonded inside the porphyrin ring of plant chlorophyll and in conjunction with light energy it produces their green colours (Fig. 2.3). Calcium features in many minerals, the most important ones biologically being fluorspar ($CaF_2$ solubility product, $K_s = 1 \cdot 7 \times 10^{-10}$ $g^3$ $ion^3$ $l.^{-3}$) which provides the natural fluoridation of our water supplies, and calcium phosphate mineral, $Ca_3(PO_4)_2$, which is believed to arise from decayed teeth, bones and sea shells. Milk and vegetables are rich sources of calcium, although the latter may have its calcium bound in an insoluble compound such as calcium oxalate. A large quantity of $Ca^{2+}$ in our drinking water comes from limestone $CaCO_3$ which, although insoluble, becomes soluble again under the action of carbon dioxide and water to give the bicarbonate

$$CaCO_3(s) + CO_2 + H_2O \rightarrow Ca(HCO_3)_2$$

Magnesium, calcium and ferrous ions in water give rise to the familiar hardness that exhibits itself when water is boiled or soap is added. Each gives precipitates and the soap precipitate of calcium stearate has been known to cause mild dermatitis.

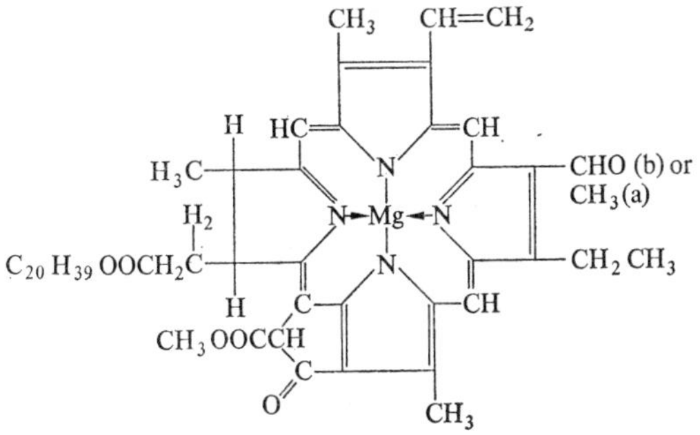

FIG. 2.3. The binding of $Mg^{2+}$ in plant chlorophylls a and b.

Magnesium ions are found complexed with nucleic acids inside cells and are necessary for nerve impulse transmissions, for muscle contractions and for the metabolism of carbohydrates. All phosphate transfer enzymes require a metal ion and magnesium is usually the one used; however, manganese may suffice instead. Overdoses of magnesium can cause depression and anaesthesia. Epsom salts $MgSO_4.7H_2O$ are used as purgatives since $Mg^{2+}$ finds it difficult to pass through the intestinal wall so it pulls water through the wall into the intestine to hydrate itself.

Humans require calcium, phosphorus and vitamin D for the formation of bones and teeth (growing children need up to $1 \cdot 5$ g calcium per day and adults $0 \cdot 5$ g). It is also needed for the formation of milk, precipitation of milk casein in the stomach, maintaining the correct rhythm of the heart beat and in the conversion of fibrinogen into fibrin to form blood clots. In fact, calcium salts are sometimes administered to hasten blood clotting or conversely, sodium or potassium citrate can be given to complex the calcium salts and so reduce the clotting. Blood donations can be taken into $3 \cdot 8\%$ potassium citrate solution. If the blood calcium level falls it is replenished by drainage from the bones and eventually osteomalacia results (skeletal bone bending). The regulation of blood calcium is controlled by the parathyroid glands using the hormone calcitonin (a chain of 32 amino acid residues, see Fig. 9.6). Administering calcitonin, isolated from parathyroid glands, may prevent osteoporosis (thin brittleness of bone) in the aged. Calcium deficiency in blood plasma causes muscular twitchings and eventually convulsions.

### 2.3 *Manganese*

Manganese is the second most abundant heavy metal ion in nature (iron being the first, of course) and it is one of the rare elements that can exist in eight different oxidation states. However, as far as biological systems are concerned, just two oxidation states are important, Mn (II) and Mn (III). Six coordinate Mn (II) is octahedral but the ion may also be five or seven coordinate (e.g. $[Mn(OH)_2EDTA]^{2-}$). Mn (II) compounds are often pale pink and have the high spin state because the $d^5$ configuration is a half-filled state and so has extra stability. The Mn (II) ion can also be tetrahedral in a non-aqueous environment. Manganese (II) chemistry resembles magnesium (II) in that the ions prefer weaker ligand donors such as carboxylate or phosphate groupings. Mg (II) can be replaced by Mn (II) in the DNA scheme and the reactions still proceed but give a different range of products.

Manganese (III) is not stable in aqueous solution unless it is complexed. The ion is easily reduced to Mn (II) in that Mn (III) is a powerful enough oxidizing agent to react with water to give Mn (II) and oxygen. When complexed, Mn (III) is quite stable (e.g. $Mn(C_2O_4)_3{}^{3-}$ the oxalate complex), and is usually found not truly octahedral but Jahn–Teller distorted.

It is known that photosynthesis cannot occur in spinach without the presence of manganese (II) and this is probably true for other plants also. Plants are our bodily source of manganese and it is needed for enzyme activation, e.g. for isocitrate dehydrogenase, malic enzyme and pyruvate decarboxylate.

## 2.4 *Iron*

Iron is the most abundant transition metal and probably the most well-known metal in biological systems. In solution it has two oxidation states, (II) and (III). Ferrous ions are less common than ferric as the ferrous can be easily oxidized up to ferric. Ferrous solutions are pale green and are difficult to keep unless acidified or complexed because even molecular oxygen can effect the oxidation. Ferric solutions are yellow or brown due to the presence of $FeOH^{2+}$. Both oxidation states have octahedral configurations. Our main bodily sources of iron lie in green leaves or meat.

Our bodies contain 4–5 g of iron, 65–70% in hemoglobin, 15% stored in the liver, spleen, marrow and kidneys and the remainder is involved in the formation of protein and in redox reactions in the plasma [2]. Colostrum and milk are low in iron content so infants require a reserve at birth and this is furnished by the mother prior to birth and is stored in the liver and spleen. This supply lasts up to six months and thereafter the body is unable to store as much iron (a small amount is stored as hemoglobin in the spleen). Excesses are excreted. Bleeding, childcarriage, parturition and menstruation require more iron than is usually found in the diet and anaemias may result unless the iron deficiency is replenished. This may be achieved by administering ferrous salts since they are easily absorbed in the intestine. Alternatively, intravenous injections may be used as iron ascorbates, citrates or colloidal carbohydrate complexes rather than the more highly ionized chloride or sulphate. Even so, minor toxic reactions are sometimes observed with the first two salts mentioned.

Depending on the the type of complexing ligand attached to an iron, the metal may be divalent (e.g. myoglobin or hemoglobin), trivalent (e.g. catalases and oxidases) or oxidized and reduced between both states (e.g. cytochrome).

The four iron (II) ions in hemoglobin (MW 65,000) are each held inside a four coordinate porphyrin ring (see Fig. 2.4) and are also attached to two of the four chains that intertwine to make up the hemoglobin molecule. This attachment is covalent to one or three imidazole nitrogens belonging to a histidine amino acid residue in these chains. The ferrous ion sits in this environment

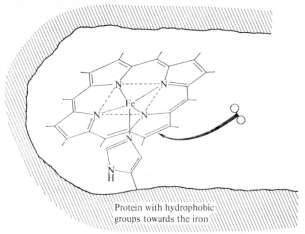

Fɪɢ. 2.4. Diagrammatic representation of five coordinate iron (II) in hemoglobin (4 bonds to a porphyrin ring and 1 to the imidazole nitrogen from a
histidine residue) being approached by an oxygen molecule.

deep inside a pocket of the protein molecule and it is this combination of complex bonds and shielding by the pocket that permits the
ferrous to be reversibly attached to oxygen and not irreversibly
oxidized up to ferric.

If we remove the cells from blood, the solution remaining is
plasma and this contains less than $0 \cdot 1 \%$ of our bodily iron. The
plasma concentration, which equals the serum concentration, shows
interesting variations with the sex, time of day and diet. The normal
male plasma has $c.$ 23 μM iron and the female 10–15 % less but this
is not really a deficiency. These values vary according to the time
of day (diurnal variation). They are highest early in the morning
and lowest in the middle of the afternoon. Diurnal variations of
between 3 and 18 μM have been observed. Nightworkers have the
converse characteristics. During pregnancy, females have an iron
deficiency because of foetal requirements and because the blood
volume expands during the later months so an increase in total
hemoglobin is necessary. Thus, from mid-term onwards, the plasma
iron drops even though the total iron binding capacity (T.I.B.C.—
see Table 1.1), which is proportional to the transferrin (MW
90,000) concentration, increases. A point to be noted here is that
anaemias cannot automatically be treated by administering iron
unless the T.I.B.C. is high or can be therapeutically raised. At birth

the infant has a higher plasma iron content (27–36 μM) than the mother but its T.I.B.C. is lower so that within a matter of hours the concentration is about 18 μM and this value then slowly rises over the next few years until it reaches adult level at the age of five years.

The average lifetime of a red cell is 120 days and 25–30 mg iron are lost each day; thus a similar quantity of iron requires to be eaten. Ingested iron, finding itself in the plasma, cannot be turned into hemoglobin without involving the marrow. On the other hand if we eat excesses of iron, siderosis conditions arise. This is frequently found among the Bantus of South Africa (they cook in iron pots!) and their females do not have an iron deficiency during pregnancy.

## 2.5 Cobalt

Cobalt has two common oxidation states that exist in solution, cobaltous (II) and cobaltic (III). Unlike ferrous, cobaltous solutions are quite stable to air oxidation. They are usually pink and have the ion octahedrally solvated. Tetrahedral configurations are also found in solution, however, e.g. $[CoCl_4]^{2-}$ and more rarely can be square or trigonal pyramidal. In the presence of nitrogen donor ligands cobalt (II) can be readily oxidized by molecular $O_2$; the first step in the reaction is believed to be the bonding of the oxygen to the cobalt (II) ion to give a transient cobalt (IV) species. In fact cobalt (II) complexes with ligands such as glycylglycine and histidine can act as reversible oxygen carriers. Complexes of cobalt (I) are known and their configurations are the same as those for cobalt (II).

Cobalt (III) is a powerful oxidizing agent and its salts decompose water. The complexes are pink, yellow, blue, violet or purple and are quite stable because they form inert complexes, e.g. $[Co(NH_3)_6]^{3+}$. Co (III) has more reported complexes than any other element.

Although cobalt is quite rare on the earth's surface, it is found in fertile soil and so in the plants thereon. In our bodies it is necessary for vitamin $B_{12}$ (see Chapter 8) (Co (III)), our richest source of this vitamin being from animal liver. $B_{12}$ or cyanocobalamin is needed for our bodies to form hemoglobin and deficiencies cause pernicious anaemia. The $B_{12}$ present in our food cannot be absorbed through the intestinal wall unless it encounters hydrochloric acid (from gastric juices) and the intrinsic factor (excreted from stomach walls) *en route* to the intestine. It is interesting that Australian sheep at one time had a $B_{12}$ deficiency because of a lack of cobalt salts in the soil. Nowadays, the deserts, low in cobalt, are sprayed with salts.

Cobalt (II) is used as an enzyme activator, e.g. in carbonic anhydrase or carboxypeptidiase. We have just listed four different possible configurations from the bonds in cobalt (II) complexes. This and other inorganic evidence indicates that cobalt (II) can enter low symmetry sites in enzymes and this is how the element is usually found.

## 2.6 *Copper*

Copper, like cobalt, is found in two different oxidation states, cuprous (I) and cupric (II). The cuprous state is easily oxidized to cupric and so is only found in solution when complexed or is found in solids, e.g. CuCl. The radius of $Cu^+$ is approximately equal to $Na^+$ and its complexes have coordination numbers of 2, linear, 3, planar, or 4, tetrahedral. Copper (I) complexes are usually colourless. The relative stabilities of the copper (I) and (II) states depends upon the dielectric constant of the surroundings and the nature of the neighbouring groups to the ion. The cupric ion is usually green, brown or blue and can be complexed with coordination number 4, tetrahedral, trigonal pyramidal, square pyramidal (e.g. as with $\beta$ alanyl L-histidine) [3] or square, or coordination number 6, distorted octahedral with four tightly held ligands in a plane and two further away on an axis (Jahn–Teller distorted as with (glycyl–L–histidinato)–Cu(II)). These fifth and sixth ligands are not strongly bound. Next to iron, cupric is chemically the best catalyst in oxidation–reduction processes.

Copper is found in enzymes, e.g. phenolase or hemocyanin (Cu (I)), and both these are capable of carrying oxygen as hemoglobin does. Copper is actually required in the production of hemoglobin. The cuproproteins use molecular oxygen as an electron acceptor and most of them contain an even number of copper ions, e.g. cerebrocuprein has 2 and is used in oxygen storage and transport in the brain, ceruloplasmin has 8 and is used in oxygen transport in blood plasma. They are usually coloured blue which is characteristic of tetragonal copper (II) and nitrogen donor ligands.

Copper is stored in the liver and eating liver or shellfish, such as oysters, provides our sources of copper. Excess copper if not excreted may be found deposited in the eyes. It has long been known that copper salts are poisonous to lower organisms and so they are used as a fungicide to destroy algae. Nevertheless, the green colouring matter of canned vegetables used to have added copper but since it has been found to be poisonous to humans when ingested in appreciable quantities, this practice is now forbidden.

## 2.7 *Zinc*

Zinc is one hundred times as abundant as copper and yet is still quite scarce. It occurs in several mineral forms, one of which is calamine $ZnCO_3$. Although both zinc and calcium occupy positions in groups II of the periodic classification their chemistries are very different because $Zn^{2+}$ is a much smaller ion. In solution the ion is coordinated with four or six ligands, tetrahedral or planar (e.g. bis(glycinyl)$Zn^{2+}$), or octahedral (e.g. $ZnCl(H_2O)_5^+$). It is noteworthy that zinc ions do not have the possibilities of crystal field stabilization energies stabilizing one configuration with respect to the others so all three configurations are possible and depend only on size considerations and electrostatic and covalent forces. As with cobalt (II), zinc (II) has the ability to occupy low symmetry sites in enzymes. In that it has a full d shell, zinc is not really a transition element but it does have the transition element properties of forming complexes and of being a strong Lewis acid. In aqueous solution there is no evidence for zinc (I) or of oxidation states higher than (II).

The body obtains zinc from plants and from other animals. It is also encountered as a filler in soft rubbers and as a white pigment in paint. Many years ago, zinc deficiency is believed to have led to dwarf formation among some peoples living in central Europe.

Zinc is an essential constituent of several enzymes, e.g. in the activator in carboxypeptidiase A (1 atom per molecule of protein). This protein has both peptidiase and esterase activity but when the zinc is replaced by cadmium or mercury, the former is removed and the latter is enhanced.

## 2.8 *Molybdenum*

Molybdenum occurs in the earth as $MoS_2$ and, as such, can be used as a solid lubricant. *In vivo* it participates in biochemical redox reactions such as xanthine and purine oxidations in milk and in the liver. It also acts as a cofactor in nitrogen fixation bacteria and as a nitrate reducer in some plants and microorganisms. During the redox reactions, Mo (V) and Mo (VI) are attached to oxygen-containing ligands. These oxidation states can only exist in solution as oxyanions or complexes but more recently evidence for III, and even IV, oxidation states has been reported and it is possible that these may just be cationic. However, in general, there is no cationic solution chemistry of molybdenum *in vivo* [4].

## 2.9 *The solvent* [5]

Water is currently considered to be an equilibrium mixture of a dense, non-hydrogen bonded fluid which contains ordered aggregates of ice-like molecules of average composition 25 $H_2O$ and these are all hydrogen bonded together. These large polymers or clusters are continually being made up or destroyed at the expense of another cluster being broken up or formed in another part of the solution. The half-life of these clusters is the order of $10^{-10}$ seconds. Adding solutes to the water naturally affects its structure. It has already been noted that cations take on water molecules as ligands of their primary hydration sphere and, in fact, these ions also have a larger, but less firmly held, outer hydration sphere. These latter waters are more at liberty to exchange with those in the bulk of the solution. Anions also are solvated and once again this solvation may be conveniently divided into two spheres. Non-ionic solutes in water, if small, can act as sources of nucleation for new clusters to form. If they are larger in size, such as some of the colloids in blood, they have hydrophilic groups on their outer surfaces and so can still be transported by having a thick solvation sphere to keep them in solution.

Seventy per cent of our body weight (BW) is water, 0·49 BW inside cells, 0·17 BW in interstitial fluid and 0·04 BW in blood plasma. Each day 1·6 l. of water is lost and imbalances from these figures soon results in severe edema or dehydration consequences.

Whole blood consists of 92% water and 8% solutes (colloids, ions, complexes). The forces of attraction between the solutes and the water exhibit themselves in the osmotic pressure of the solution. A solution that is isotonic with blood is 0·15 M sodium chloride at pH 7·4. This solution exerts the same resultant osmotic pressure and so may be assumed to have the same need for water for solvation purposes. A 92% aqueous solution is approximately 50 M in $H_2O$. This gives roughly 100–200 $H_2O$ per ion. In addition waters are attracted to the walls of the blood vessels. Hence there appears to be precious little water and space left for cluster formation. It appears that all the water available in blood is earmarked for solvation purposes or is being transported to a position where water is required.

Finally, salts that are not fully ionically dissociated in solution may be ion paired, e.g. $[Co(histamine)_3]^{3+},Cl^-$, where the ion pair bond is roughly one ten thousandth of the strength of a normal covalent bond. The chloride ion is well known for its ion pairing capabilities and in general the strengths of ion pair bonds lie in the order $SO_4^= > Cl^- > Br^- > I^- > NO_3^- > ClO_4^-$. Hence, any complexing reaction in solution must occur through this sea of ion paired and highly solvated complexes and ions and of ordered solvent. Before any new bond can be formed many molecules of solvent must be released or reorganized.

## REFERENCES

[1] MATHEWS, B. F., *Chemical Exchanges in Man*, p. 61 (Oliver and Boyd, Edinburgh, 1967).
[2] RAMSAY, W. N. M., *Adv. Clin. Chem.*, **1**, 1 (1958).
[3] FREEMAN, H. C. *et al.*, *Chem. Comm.* 598 (1965); 23 (1966).
[4] SPENCE, J. T., *Co-ord. Chem. Rev.*, **4**, 475 (1969).
[5] FRANKS, F. and IVES, D. J. G., *Quart Rev.* 1 (1966); IVES, D. J. G. and LEMON, T. H., *R.I.C. Rev.* 62 (1968); ERLANDER, S. R., *Sci. J.*, **11**, 60 (1969).

C

## FURTHER READING

GRILLOT, G. F., *A Chemical Background to Nursing and other Paramedical Programs* (Harper International, March, 1966).

COTTON, F. A. and WILKINSON, G., *Advanced Inorganic Chemistry*, 2nd edn (Interscience, New York, 1966).

MAHLER, H. R. and CORDES, E. H., *Biological Chemistry* (Harper International, 1966).

*Chapter 3*

# LIGANDS *IN VIVO*

---

METAL IONS in biological systems are being examined from the point of view of complexing reactions. The metal ions have been considered, next we consider the ligands.

## 3.1 *Amino acids*

Living organisms contain millions of proteins having an incredible range of functions and yet these proteins are made of just 26 simple building blocks—the amino acids. As with other types of biopolymer, the proteins are *linear* combinations of the simple units, in this case, of amino acids. Tissue is always in a state of being constructed or destroyed and so free amino acids are always present in the blood stream. Nature is so arranged that these amino acids can be used over and over again. The common amino acids are listed in Fig. 3.1 and contain a carboxylic acid grouping and an α amino group (this is usually a primary amine but see Pro and Hypro). At physiological pHs the amine group is usually protonated and the carboxylic acid grouping is ionized:

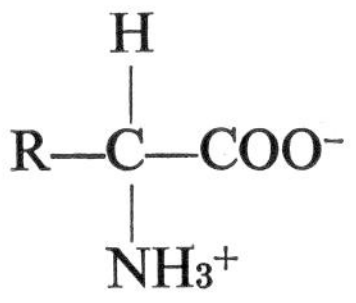

$$
\begin{array}{c}
H \\
| \\
R\!-\!C\!-\!COO^- \\
| \\
NH_3^+
\end{array}
$$

Both the amine and carboxylate groups are capable of reacting with metal ions as indeed are many other groups present in the side-chain, e.g. —OH in Ser, or Tyr, another —COO⁻ as in Asp

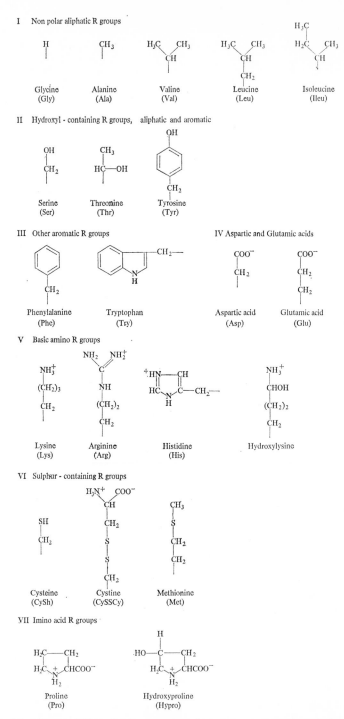

FIG. 3.1. Names (abbreviated names) and structural formulae of the side-group R in common amino acids. Pro and Hypro are shown in full. This compilation is based on that of Steiner [1].

or Glu, —SH, or —S—S— as in CySH or CySSCy, and another amine as in Arg or His.

With the exception of glycine, all the amino acids found in protein are optically active and have the L configuration (c.f. glyceraldehyde). The direction and magnitude of the optical rotation depend upon the type of side-chain and the pH of the solution. D amino acids are found *in vivo*, however, in bacterial cell walls and among many administered antibiotics, e.g. gramicidin S. In later chapters the energies of bonding to metals will be discussed and whether we are dealing with L or D is of little consequence as the heats and entropies of bonding are usually the same as long as we work with all L or all D ligands. On the other hand, enzymes are stereospecific in that an enzyme designed to take an L substrate will not usually accept a D amino acid.

Because amino acids have a positive and a negative charge on their zwitterionic forms, they are highly soluble in water and only sparingly soluble in organic solvents. The charges also give them high melting points. Chemically they are all similar in that each amino acid will exhibit an amine reaction, a carboxylate group reaction, or a reaction, such as chelation, in which both of these groups are working in unison:

$$R-\overset{\overset{\displaystyle H}{|}}{\underset{\underset{\displaystyle H_2N:\,\nearrow}{|}}{C}}-\overset{\overset{\displaystyle O}{\|}}{C}\underset{M^{n+}}{\overset{\displaystyle }{\diagdown}}O^-$$

The differences between the amino acids lie in their side-chains (R). Some, such as histidine, have the possibilities of being tridentate to a metal ion by having an extra donor group in R. Others have different shapes and sizes of side-chains and these (i) introduce steric factors into the normal amino acid chelation just pictured, and (ii) are of major importance in dictating the shapes of peptide chains, e.g. proline usually introduces a bend in the chain.

Not all the amino acids listed are essential to our diet because our bodies have the ability to synthesize some amino acids from others. (This process involves proteins as catalysts). The essential amino acids are arginine, histidine, isoleucine, leucine, lysine, methionine, phenylalanine, threonine, tryptophan and valine. Milk is an excellent source of these since the lactalbumin therein contains them all. Amino acids are decomposed in the liver to give urea,

$$NH_2—\underset{\underset{O}{\|}}{C}—NH_2$$

and this is carried by the blood to the kidneys and then it is excreted.

### 3.2 Amino acid derivatives

The next level of sophistication built upon amino acid foundations is amino acid derivatives which fall into two groups— ordinary derivatives and pedigree derivatives (i.e. amino acid derivatives of amino acids ≡ peptides). A few examples of ordinary derivatives are listed in Table 3.1. They are frequently hormones or chemotherapeutical reagents.

TABLE 3.1. Frequently encountered amino acid derivatives

| Name | Formula | Notes |
|---|---|---|
| Histamine | | The decarboxylation product of histidine. It is a vasodepressor. Skin rashes arise when histamine is present in excess. |
| Adrenaline | | It is derived from tyrosine and is a hormone that elevates blood glucose by calling upon glycogen reserves. |
| Serotonin | | A derivative of tryptophan. It is a vasoconstrictor and is used in nerve impulse transmission. |
| Penicillamine | | A derivative of cysteine and a component of antibiotics. |
| Monosodium L Glutamate | | Causes Chinese restaurant syndrome (headaches, burning sensations, face and chest pressures). Threshold = 2 g per meal. |

## 3.3 *Peptides*

These are formed by eliminating a water molecule from two or more amino acids (as $OH^-$ from $COOH$ and $H^+$ from $NH_2$). Several such polymerizations give polypeptide chains as shown in Fig. 3.2. These molecules are named after their constituent amino

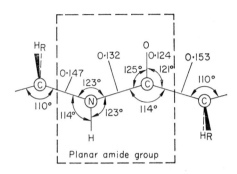

FIG. 3.2. Peptide bonds showing the position of the side-chains, the bond angles and distances (in nanometres). Taken from Steiner [1].

acid residues, e.g. glycylhistidylglycine is a combination of a glycine, a histidine and another glycine by eliminating $2H_2O$. The term polypeptide applies to simple peptides, such as the three amino acid residue one just mentioned, up to those containing very many amino acids and having molecular weights of several thousands. Thereafter the polymers are called proteins. The threshold between polypeptides and proteins is arbitrarily set at MW 5000 since the smallest physiologically active polymers, the protamines (found in spermatozoa), have molecular weights commencing at this value.

Polypeptide chains are never really found as straight chains as (a) the —N—C—C—N—C pattern is puckered, (b) the chain may be twisted about itself because of the steric influences of the side-groups R' and R", etc., and (c) as well as the peptide bond just discussed, other links can occur between amino acid residues on the same chain or between adjacent chains. The chemical bonds involved in these other links are depicted in Fig. 3.3.

Thus, the groups of amino acids that could donate electrons to a proton or a metal ion are now extended to include a peptide bond

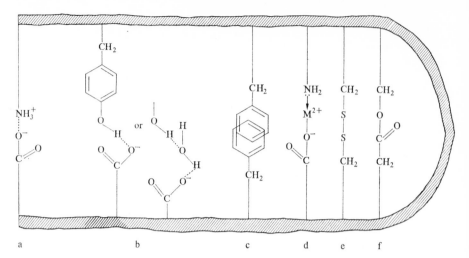

Fig. 3.3. Types of cross-linking between peptide chains. a, electrostatic bond; b, hydrogen bonds; c, hydrophobic bond; d, complex bond; e, disulphide bridge; and f, ester linkage.

nitrogen and a disulphide bridge. Further, since the chains may be puckered, it is possible for several different groups to be near enough to a metal ion to complex at any one time.

Slight variations may be found in peptides that have the same function (a) between different species, e.g. insulin, needed for the metabolic control of blood glucose, has a serine in man where there is a glycine in horse, and (b) by mutational errors in the same species, e.g. sickle cell anaemia arises because a glutamic acid residue in hemoglobin A is replaced with a valine residue.

### 3.4 *Proteins*

Their role may be divided into work proteins (transport, digestion, etc.)—this work is controlled by nucleic acid systems which have the power to pass on information—and catalytic proteins (enzymes). These enzymes do no work themselves but persuade other systems to work faster. They have molecular weights from 5000 up to many million and because they are only catalysts they must have a means of regenerating themselves, i.e. the enzyme must *not* finish up destroyed at the end of the reaction. Although an enzyme is just one variety of protein, this power of regeneration applies to many of the other types also. This is achieved by a finesse in the molecular geometry of the protein that requires four levels of structure.

1. Primary structure: This is the unique sequence of amino acid residues that varies from protein to protein.
2. Secondary structure: This is the pattern into which the chains are bent.

As already mentioned, the sharpest bends occur if a proline group is present and if the chains happen to bend back upon themselves they can be bound there by a variety of intra-chain bonds (see Fig. 3.3).

3. Tertiary structure. This is the way in which chains mutually interweave around and through other chains, each stabilizing the other. Together, two, three or four of these chains combine to give the overall shape of the protein such as the long thin strands of fibrous proteins and the compact pellets of globular proteins such as hemoglobin. It ought to be noted here that not all the reactive side-groups in a protein are in a position to be utilized in intra- or inter-chain bonding or in bonding to a metal ion. Those on the inside of a protein moiety are hindered by neighbouring chains. Thus it is only those amino acid residues around the outer edges of a protein and possibly those lining a cleft that can chemically react. Nevertheless, most of the side-groups help in setting up the correct electrostatic environment for enzyme activity. Clearly, if we have a water soluble protein it must have water attracting outer groups such as —OH or —COO⁻.

4. Quaternary structure. This is the way in which one protein attaches itself to another to form shapes familiar and visible to us, e.g. a combination of keratins gives hair.

How ought the role of proteins be defined? Briefly one might answer, 'to supply life'. The chemical concentrations of substances in living matter often greatly exceed those expected from truly thermodynamic steady state considerations. This is because the reactions governing the equilibria are coupled with protein reactions that continually supply energy to prevent the claims of thermodynamics. We have already seen an example of this in the alkali metal 'ion pump'. Steiner has related the death of an organism to be the 'reassertation of the claims of thermodynamics, corresponding to a transition from a thermodynamically un-favoured to a favoured state' [1]. We must take care to consider both parts of these coupled reactions. A more pedantic answer to the question can be conveniently given by dividing the many millions of jobs done by proteins into four classes, enzymes, hormones, nucleoproteins, and blood proteins.

3.4.1 *Enzymes* These are defined as proteins with catalytic properties. These properties are not over the surface of the enzyme but localized in an 'active site'. In fact, only a small proportion of the amino acid residues actually line this active site. The remainder are involved in holding these chosen few in the correct three-dimensional positions and in creating the correct electrostatic environment around the site. From the nature of the above definition it could be assumed that we could live without enzymes, but only very, very slowly. Enzymes work by forming a transient complex and the mechanism often involves metal ions. These may either be strongly bound and an integral part of the active site or they may have weak and reversible binding to the site. Enzymes are very specific both for the metal ion and for the substrate and an enzyme that catalyses one reaction is usually incapable of catalysing another. Most of us have warmed a reaction flask to persuade a reaction to occur more quickly. We cannot locally warm our bodies to catalyse a reaction but never-theless enzymes are capable of concentrating the free energy just in the correct position to achieve an enhanced biological reaction. Speaking through analogies, it is not always convenient to have an unguarded electrical circuit lying around when not actually required. Similarly, it is inconvenient to have these energy rich active sites unguarded when not required so nature has provided them with a guard or insulator. This is called an inhibitor and is usually a small polypeptide that sits in the active site. The inhibitor plus enzyme is known as a precursor and after removing the inhibitor the pattern of the active site is just the correct shape to receive the substrate. On the other hand, some enzymes are inactive (called proenzymes or zymogens) because they are short of a portion (called the activator), e.g. stomach hydrochloric acid is the activator for pepsinogen. Even so, some enzymes are still not quite right (now called apoenzymes) and these need yet another activator known as a coenzyme (these are often vitamins), e.g. peroxidase $C_{34}H_{32}O_4N_4Fe(III)OH$ = hematin (the low molecular weight coenzyme that contains iron) + apoenzyme (the major part of the enzyme; it is just a bulk of protein). Finally, there must, of course, be enzyme prohibitors (called antienzymes) which prevent enzyme activity when sufficient has occurred, i.e. they inactivate the active sites, e.g. the cell walls of the stomach and intestine contain antienzymes so that the walls are protected against the digestive enzymes. This proliferation of terms must not be allowed to cloud the fact that it is the fitting of the substrate into these 100% perfect active sites that is the secret of success of enzyme activity. If the pH is changed some protons may ionize from these sites, or if we expand the active sites by warming, the enzyme no longer fits the substrate (see Fig. 3.4) and so the enzyme is inactive.

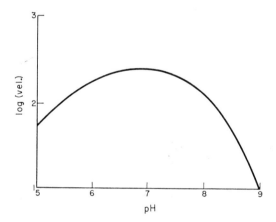

FIG. 3.4. An example of the pH dependence of the velocity of an enzyme reaction (fumarase). From Steiner [1]. Note that log (velocity) is plotted.

If irreversible damage has been done to it, it is said to be denaturized. The electrical circuit is quite a useful analogy since electrons are often transferred to or from the enzyme site during the catalysis reaction.

3.4.2 *Hormones* These are another group of control substances,

closely related to enzymes. They function at extremely low concentrations sometimes by interacting with an important enzyme
and sometimes by altering the permeability of membranes. The
three major subgroups of hormones are amino acid derivatives,
peptides or proteins, and steroids. As their concentrations are so
low they will not be considered further here but examples of each
of these three varieties of hormones are given in Fig. 3.5.

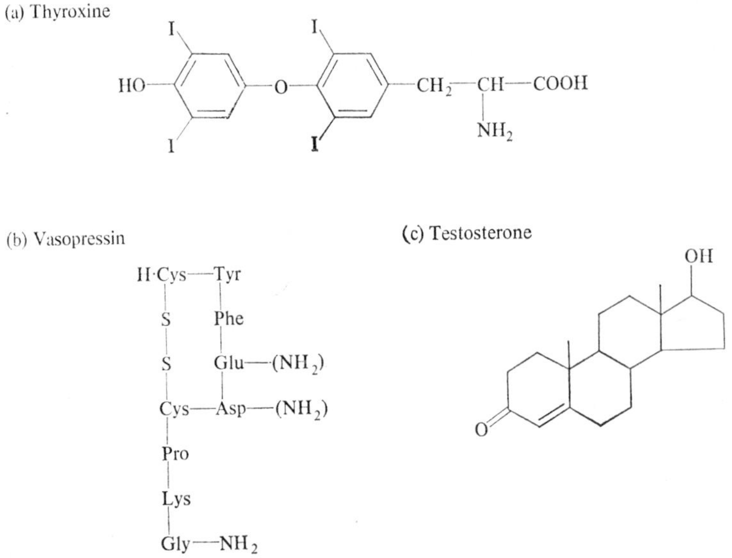

(a) Thyroxine

(b) Vasopressin

(c) Testosterone

FIG. 3.5. Examples of the three different varieties of hormones. (a) amino acid
derivatives; (b) peptides or proteins; and (c) steroids.

3.4.3 *Nucleoproteins* (MW $= 10^8$ or higher). These are combinations of nucleic acids (see later) and protein and are required so
that organisms can reproduce themselves in faithful likenesses.
There are two varieties of nucleic acids—ribonucleic acids (RNA)
and deoxyribonucleic acids (DNA). Both are linear polymers of
nucleotides. The DNA protein occurs mainly in the nuclei and the
RNA in the cytoplasm of cells. The DNA protein possesses the
necessary genetic information; the RNA meets the DNA and
memorizes its information and then the RNA directs the construction of peptides from an assortment of amino acids. Once

again perfect fitting of protein into peptides is a key factor in their mechanism. Nucleic acids tend to be negatively charged and proteins positively charged so that the nucleic acid protein complex is held together by mainly electrostatic interactions.

3.4.4 *Blood proteins* The blood stream is rich in proteins. We have the already-mentioned enzymes and hormones and there are also the albumins (which constitute more than half the blood protein) and the globulins. Although continuously changing, the average composition of the blood remains fairly constant. The albumins keep the osmotic pressure constant and help to buffer the pH at 7·4. Meanwhile, the globulins are concerned with transport. These globulins are usually insoluble in water but soluble in dilute salt solutions. The $\alpha$ and $\beta$ globulins originate in the liver and are concerned with transport and blood clotting. The $\gamma$ globulins play the important role of conferring immunity by producing antibodies which complex with foreign substances.

## 3.5 *Nucleic acids and nucleotides*

Fig. 3.6 shows the nucleic acids as polymers of 5-carbon sugars and phosphate groupings. The phosphatediester bonds join the 3′ and 5′ carbons of adjacent sugars. Each sugar has a basic heterocyclic nitrogen ring at the 1′ position. The sugars are 2-deoxy-D-ribose for DNA and D-ribose for RNA. The phosphate group

$$
\begin{array}{c}
O \\
\parallel \\
-O-P-O- \\
| \\
OH
\end{array}
$$

usually has just one ionizable proton free and at neutral pH this is ionized to give a negatively charged polymer (clearly an ideal ligand for a metal ion). The phosphate groups at the end of chains have two ionizable protons. There are five varieties of side-group bases frequently encountered as the side-groups of these polymer chains: adenine and guanine (the purines) and uracil, cytosine and thymine (the pyrimidines). Both varieties are rich in nitrogen donor atoms.

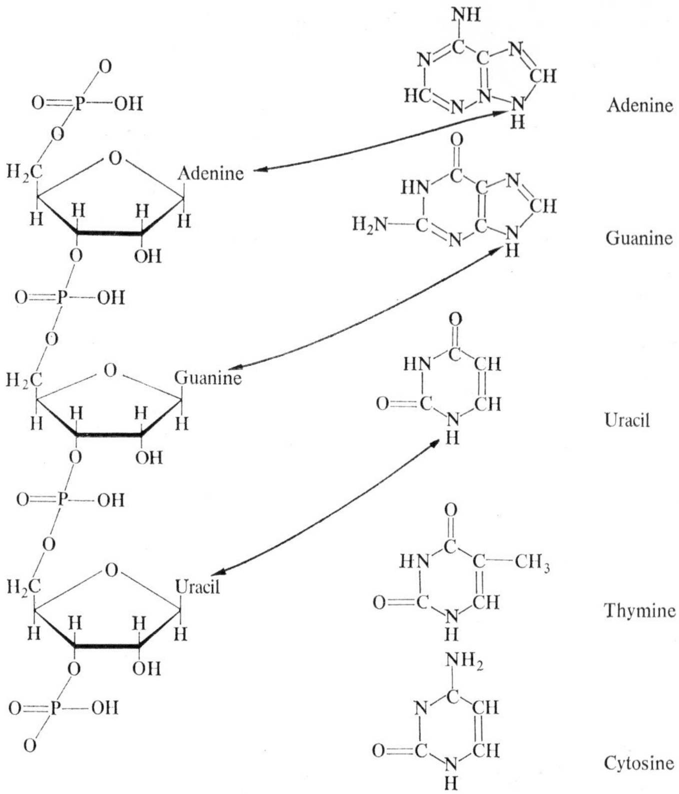

FIG. 3.6. Basic structure of ribonucleic acids. Taken from Steiner [1].

## 3.6 Carbohydrates, carboxylic acids and lipids

During the process of digestion, polysaccharides are split down into monosaccharides (e.g. fructose or glucose, see Fig. 3.7) and these are carried by the blood to the liver where they are rebuilt into a variety of polysaccharide that can be stored therein (glycogen). The conversion of glycogen into energy is a complicated process that liberates carboxylic acids as biproducts, e.g. pyruvic $CH_3.CO.COOH$ or lactic acids $CH_3—CHOH—COOH$. The carbohydrates have primary and secondary alcohol groups and keto groups that can act as possible donors of electrons to metal ions. In addition to some of these groups, the carboxylic acids have their acid groupings that at blood pHs will be in the —COO⁻ form. The carbohydrates are important in that they act in both a

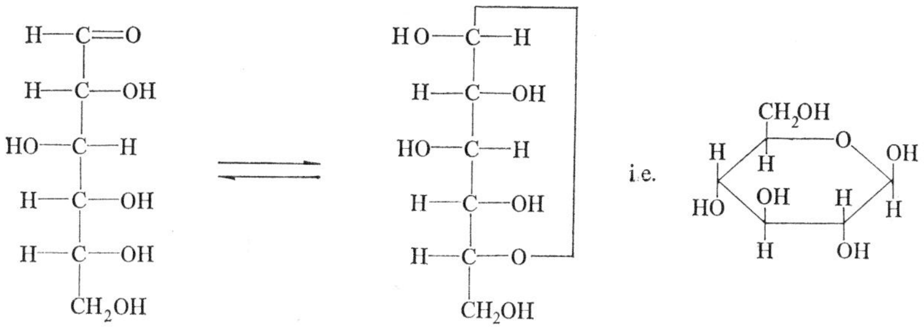

FIG. 3.7. Glucose is in equilibrium between straight chain and six membered ring forms.

structural role and as a means of energy storage. They can appear as a five- or a six-membered ring and as a chair or boat form. From an entropy viewpoint, the interactions of these different forms of the same chemical with a metal ion would be extremely interesting. Unfortunately they are sometimes slow to equilibrate and also they oxidize in the presence of the metal ion.

The term 'lipid' covers a miscellany of plant and animal chemicals that have just one feature in common—they are insoluble in water. The following is a brief list of the different types of lipids (see Fig. 3.8).

1. Triglyceride fats and fatty acids. Figure 3.8 shows the fats to be esters of glycerol (mono, di or tri esters giving mono, di or triglycerides) with linear monocarboxylic acids known as the fatty acids. The triglycerides are used as fuel and nutritional reserves. The fatty acids vary in their solubilities: the longer the chain length, the lower the solubility; the more double bonds there are present, the higher the solubility.

2. Phosphatides are diesters of phosphoric acid. They are hydrophobic and are important in nerves as an electric insulating part of membranes.

3. The terpenes include the carotenoids (linear chains of single and double bonds) which include vitamin A. This is needed for night vision.

4. The steroids are hydrocarbon side-chains attached to four fused rings (three six membered and one five membered), e.g. cholesterol. This is found in all tissues in the structural parts of membranes.

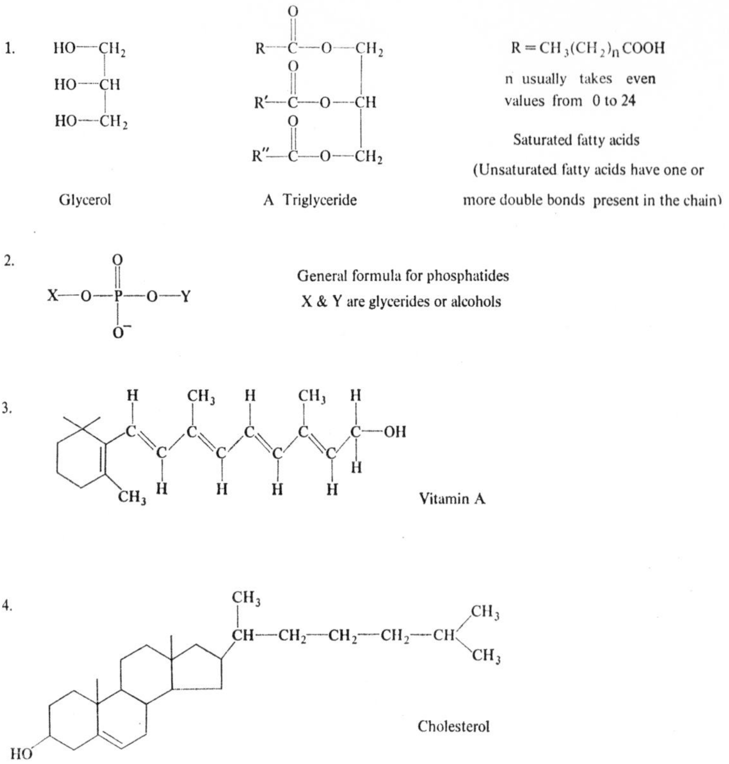

FIG. 3.8. Four varieties of lipids: 1. Triglyceride fats. 2. Phosphatides. 3. Terpenes. 4. Steroids.

Cholesterol precipitation in the walls of blood vessels gives atherosclerosis, and in the gall bladder gives gall stones.

The low solubility of lipids has restricted researches into metal ion–lipid complexes.

### 3.7 *Simple anions*

The role of water as a ligand has already been mentioned in Chapter 2. These complexed waters can easily lose a proton at alkaline pHs to give a hydrolysis product that contains the hydroxide ion as a ligand.

$$M(H_2O)_x^{2+} + OH^- \rightleftharpoons M(H_2O)_{x-1}OH^+ + H_2O$$

In general, the higher the pH, the greater the tendency to lose the proton from an already complexed water molecule. At lower pHs the mechanism is one of replacing a water molecule with a hydroxide ion. There is reason to believe that both these mechanisms occur around the blood pH (7·4). The solubility products and approximate pHs at which 10 mM solutions of metal ions precipitate as hydroxides are given in Table 3.2. Even if we change

TABLE 3.2. Solubility products and approximate pHs at which 10 mM metal ion solutions first precipitate [2], [3]

| Metal ion (10 mM) | Solubility product (g ion $1^{-1})^n$ | Approx. pH of onset of precipitation |
|---|---|---|
| Mg(II) | $3·1 \times 10^{-11}$ | 10 |
| Ca(II) | $1·3 \times 10^{-6}$ | 12 |
| Mn(II) | $2 \times 10^{-13}$ | 9 |
| Fe(II) | $1·8 \times 10^{-15} - 8 \times 10^{-6}$ | — |
| Fe(III) | $6 \times 10^{-38}$ | 2 |
| Co(II) | $2·5 \times 10^{-16}$ | 8 |
| Cu(II) | $2 \times 10^{-19}$ | 6 |
| Zn(II) | $1 \times 10^{-17}$ | 6 |

to different metal ion concentrations and to another solvent, such as blood, the same order is expected to persist. Clearly, cupric and zinc ions have to be highly complexed in blood or else present in so small concentrations that precipitation does not occur.

Other anionic ligands found *in vivo* include $CO_3^=$, $HCO_3^-$, $PO_4^{3-}$, $HPO_4^=$, $H_2PO_4^-$, $SO_4^=$, $F^-$, $Cl^-$, $Br^-$ and $I^-$. Phosphate, sulphate and bicarbonate are the main ions inside cells, whereas chloride and bicarbonate are the ones outside. These ions are as essential to the proper functioning of the human bodies as are the cations of Chapter 2. The sodium salts of the chloride and iodide occur in our normal diets. The chloride provides the hydrochloric acid for the stomach and the iodine is needed in thyroglobulin. Deficiencies produce hypothyroidism and goitre. Mercurous chloride, $Hg_2Cl_2$, may be used as a cathartic and mercuric chloride, $HgCl_2$, as an antiseptic. Sodium, potassium or ammonium bromide may be used as a sedative. Fluoride is necessary for the correct construction of the matrix of bones and teeth; deficiencies lead to tooth decay, overdoses lead to mottling.

D

The optimum is taken as 1 p.p.m. in one's daily drinking water. However, some foods and drinks contain much higher concentrations, e.g. canned fish 12 p.p.m, tea 5–7 p.p.m. The carbonates and phosphates are used as pH buffers and arise from the metabolism of carbohydrates and nucleic acids. The rate of deposition of phosphate in the skeleton depends upon the $Ca^{2+}$ and vitamin D concentration. $Ca_3(PO_4)_2$ is an important constituent of bones. Carbohydrate metabolism and the glycogen storage system also require phosphates.

### 3.8 *Chemotherapeutical reagents*

It is often surprising to count the number of artificially introduced ligands that we have within us at any one time. These are listed in full in many biochemical or medical textbooks but a few are listed in Fig. 3.9 to illustrate the wide variety of different donor groups that we subject ourselves to.

*Saccharin* or nowadays sucaryl (a cyclamate) is used as a sweetening agent having no food value. It is recommended for diabetics as it contains no sugar.

*Aspirin* and *phenacetin* are pain and fever reducing (analgesic and antipyretic) agents.

*Sulphanilamides* or sulphapyridines are used to cure streptococcus infections, gonorrhoea and pneumonia.

*Novocaine* or cocaine are local anaesthetics and are administered by subcutaneous injection. Cocaine is habit forming, novocaine is not.

*Alkaloids* all contain nitrogen heterocyclic rings. They include strychnine, morphine, codeine, heroin, quinine, nicotine and lysergic acid diethylamide. They can be used as specific stimulants. Excesses are toxic.

*Barbitals* are also heterocyclic nitrogen rings derived from barbituric acid. They are used as hypnotics and sedatives. They are cumulative poisons.

*Antibiotics* such as the penicillins (Table 3.1) or the streptomycins are extracted from moulds and enable us to control most common ailments. They function by preventing bacterial growth or by killing the bacteria entirely.

*Simple organic molecules.* Quite simple molecules can sometimes

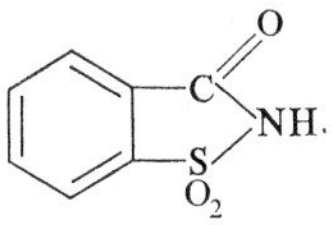

Saccharin

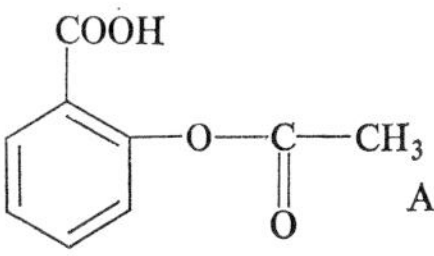

Aspirin

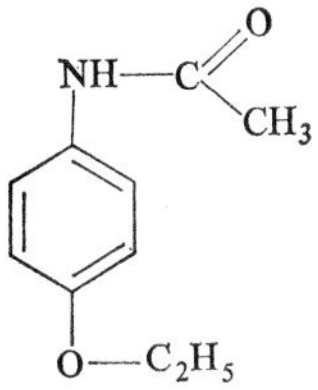

Phenacetin

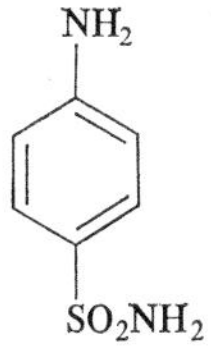

Sulphanilamide

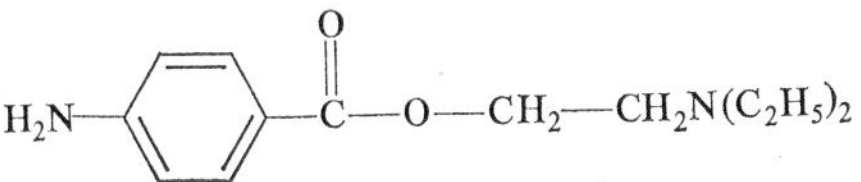

Novocaine

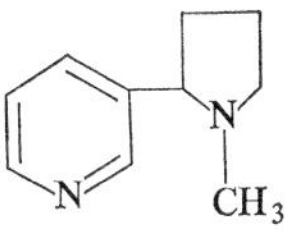

Nicotine

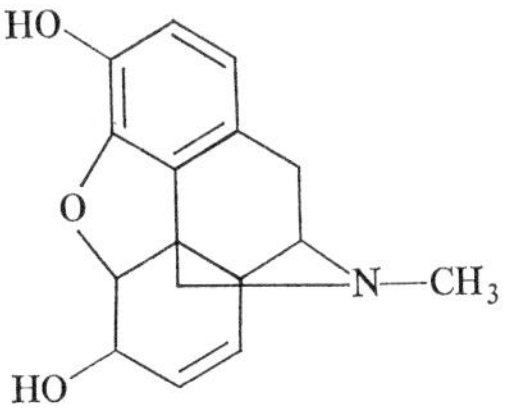

Morphine

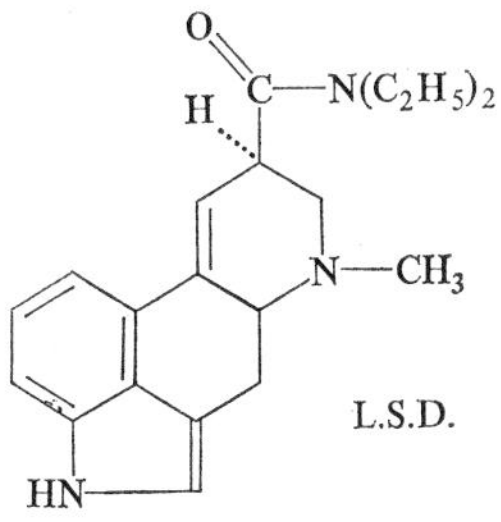

L.S.D.

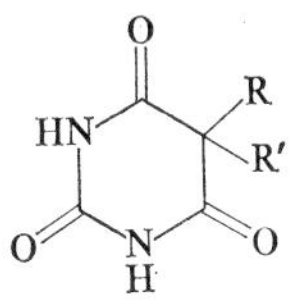

General Barbital formula

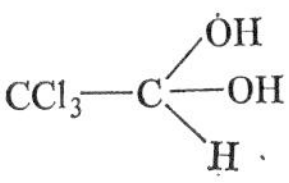

Chloral hydrate

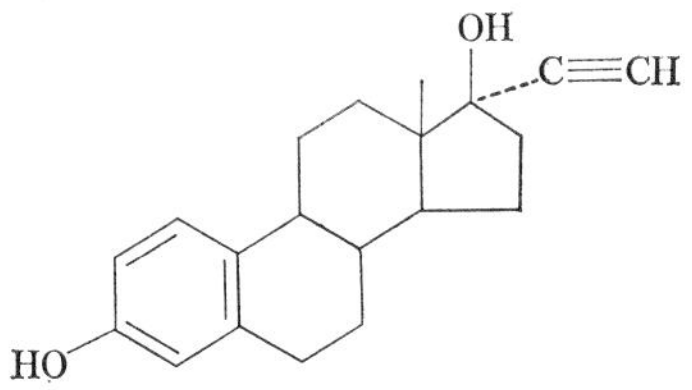

17 α - ethinyloestradiol

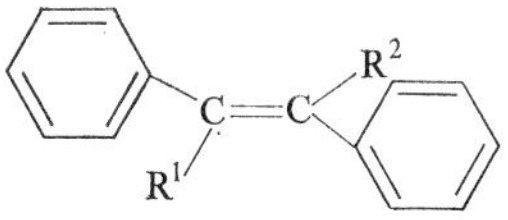

General substituted diphenylethylene formula. (Antizygotic reagents)

FIG. 3.9. Artificial ligands found *in vivo*.

have a profound effect upon our nervous systems. Chloral hydrate is a well-known hypnotic and ethyl alcohol is so dangerous that the British government has restricted its blood concentration to less than 17 mM for driving a motor vehicle. 110 mM is usually lethal.

*Chemical contraceptive pills.* In Britain the pill usually consists of a mixture of *synthetic* oestrogen and progestin. The *naturally* occurring oestrogens are not effective when taken orally. They are taken for 21–25 days per month. Their action is one of pseudo-pregnancy. In Sweden, the 'morning after' pill F6103 is now available. It works by an antizygotic mechanism that antagonizes the progesterone and so destroys the corpus luteum.

The author has not, by any means, listed all the possible ligands in human bodies and yet sufficient examples have been given to indicate the types of donor groups that will be encountered in the active centres mentioned in later chapters.

## REFERENCES

[1] STEINER, R. F., *Life Chemistry* (Van Nostrand, U.S.A., 1968).
[2] GURD, F. R. and WILCOX, P. E., *Adv. in Protein Chem.*, **2**, 311.
[3] FEITKNECHT, W. and SCHINDLER, P., 'Solubility Constants of Metal Oxides, Metal Hydroxides and Metal Hydroxide Salts in Aqueous Solution', *Pure and Applied Chemistry*, **6**, 130 (1963).

## FURTHER READING

MAHLER, H. R. and CORDES, E. H., *Biological Chemistry* (Harper International, 1967).
GRILLOT, G. F., *A Chemical Background to Nursing and other Paramedical programs* (Harper International, 1966).
TAYLOR, R. J., *The Chemistry of Proteins* (Unilever Educ. Booklet, 1969).

*Chapter 4*

# WHICH METAL IONS REACT WITH WHICH LIGANDS?

CHAPTERS 2 and 3 list a range of metal ions and a range of ligands. Studies would be considerably enhanced if one knew which ligands formed the strongest bonds to each of the metal ions, but this would involve knowing the values of the bond strengths for many millions of bonds. It is doubtful if 1 % of these values are at present available and so we have to resort to less rigorous methods of estimating bond strengths. The bond strengths, be they measured or estimated, must try to answer two questions.

1. Why do some metal ions prefer one type of ligand and others another? Or conversely, can we predict which groups on an enzyme will be most likely to be attached to a given metal ion?
2. If, for medical reasons, one wishes to remove a ligand from a metal ion to which it is complexed or if we wish to extract an excess of one particular type of free metal ion from the blood, how do we choose the correct competing ligand or sequestering agent?

The answers lie in the various theories of bonding, of which there are many. Unfortunately, the more quantitative and precise the predictions of any theory, the narrower the field of chemistry to which this theory has been applied. This narrowing arises because the more exact the theory, the greater the number of assumptions (or parameters assumed or held constant) in its derivation. Several years ago, the chemists referred to in the references at the end of this chapter attempted the converse approach, i.e. to develop a theory that would apply to all branches and states of chemistry. Clearly such a theory would initially be only semiquantitative and would need to use the minimum number of assumptions.

### 4.1 *The principle of hard and soft acids and bases* (*HSAB*)

This approach assumes only:

1. that if a bond exists between two atoms, one will play the role of an acid and the other a base, and
2. that electrons hold the bonded atoms together.

Electronegativities, the shapes of orbitals and the wave equations have all been set aside for the present. The (Lewis) acid is taken as the species (atom, molecule or ion) that has vacant accommodation for electrons and the base has the tendency to give up electrons. A typical acid-base reaction is:

$$A + :B \rightarrow A : B$$

This A—B bond may be any chemical bond, e.g. $[(H_3N)_5Co—NH_3]^{3+}$, $C_2H_5—OH$, $C_2H_5O—H$ or $H—OH$. Thus we mentally dissect any species into an acid and a base fragment. We need not consider whether the fragments could exist in isolation; this is usually irrelevant. Two further points may be noted here: (a) when A is a metal ion :B may be called a ligand, and (b) when a salt dissolves in water, the cationic portion becomes bonded to the basic end of water molecules and the anionic to the acid end, i.e. $A(:OH_2)_n$ and $(HOH)_m:B$.

#### 4.1.1 *Acid and base strength*

The strengths associated with the acid or base at the ends of a bond are said to arise from two factors, their intrinsic strengths (*S*) and their 'softness' parameters ($\sigma$). Then the strength of the A—B bond, which as a first approximation is proportional to its formation constant *K*, is defined as:

$$\log K = S_A.S_B + \sigma_A.\sigma_B$$

'Softness' arises from the electron mobility or polarizability of a species. If the electrons are easily moved, the species is soft; if firmly held, the species is hard. Other descriptions of a soft *base* would include such terms as highly polarizable, easily oxidized or valence electrons loosely held. Clearly this is all associated with a low density of charge on the base. A hard *acid* is of small size, high positive charge density and usually does not contain unshared pairs of electrons in its valence shell. Naturally, a hard base and a soft acid are the converse of these descriptions. Table 4.1 summarizes these definitions.

4.1.2 *Classification of acids and bases*   The single principle behind HSAB is: a strong bond is formed by a hard acid combining with a hard base or a soft acid with a soft base. Hard–soft bonds are weak.

We are now in a position to divide all our species into hard or soft acids or bases either by asking the questions 'Is this species polarizable?' 'Does it have a large positive charge density?' or, more quantitatively, by measuring log $K$ values for a series of compounds related via a common acid or base, e.g. we might measure the log $K$ (in this case the pK) values for H—F, H—Cl, H—Br, and H—I. Here the standard $H^+$ is used as a reference point for the four halide bases. If the polarizability of the proton is set at 0 ($\sigma_A = 0$)—it has no electrons—and the intrinsic strength factor of the proton is arbitrarily set at 1 ($S_A = 1$), we can calculate $S_B$ for $F^-$, $Cl^-$, $Br^-$, and $I^-$. Similarly we can take a standard base, e.g. $F^-$, and use it as a reference point to place a selection of metal ions in hardness order. In fact, Tables 4.2 and 4.3 have been compiled by just this means. In practice it is easy to compile these tables from a book of stability constants as the hard acids prefer ligand donor atoms of the 1st short period

$$N \gg P > As > Sb$$
$$O \gg S \sim Se > Te$$
$$F \gg Cl > Br > I$$

and the soft acids prefer thus

$$N \ll P > As > Sb$$
$$O \ll S < Se \sim Te$$
$$F \ll Cl < Br < I$$

e.g. log $K_{CuI} >$ log $K_{CuF}$ ∴ $Cu^+$ is soft.

4.1.3 *Uses of HSAB*   Thousands of examples of the HSAB idea that strong bonds are only formed between hard–hard or soft–soft components have been recorded and many are listed in the

TABLE 4.1. Hard and soft acids and bases: Some *properties* that can be used as guide lines for classifying species. The whole column need not be true before a species is called hard or soft but the more factors that are true, the greater the degree of hardness or softness

| ACID (Electron acceptor) | | | BASE (Electron donor) | | |
|---|---|---|---|---|---|
| Property | Hard | Soft | Property | Hard | Soft |
| Polarizability | Low | High | Polarizability | Low | High |
| Electropositivity | High | Low | Electronegativity | High | Low |
| Positive charge or oxidation state | Large | Small | Negative charge | Large | Small |
| Size | Small | Large | Size | Small | Large |
| Types of bond usually associated with the acid | Ionic, electrostatic | Covalent, π | Types of bond usually associated with the base | Ionic, electrostatic | Covalent, π |
| Outer electrons on donor atoms | Few and not easily excited | Several, easily excited | Available empty orbitals or donor atom | High energy and inaccessible | Low lying and accessible |

TABLE 4.2.   HSAB Classification of acids from PEARSON, R. G., *J. chem. Educ.*, **45**, 581, and 643 (1963)

| Hard | Soft |
|------|------|
| $H^+$, $Li^+$, $Na^+$, $K^+$ | $Cu^+$, $Ag^+$, $Au^+$, $Tl^+$, $Hg^+$ |
| $Be^{2+}$, $Mg^{2+}$, $Ca^{2+}$, $Sr^{2+}$, $Mn^{2+}$ | $Pd^{2+}$, $Cd^{2+}$, $Pt^{2+}$, $Hg^{2+}$, $CH_3Hg^+$, |
| | $\quad Co(CN)_5^{2-}$, $Pt^{4+}$, $Te^{4+}$ |
| $Al^{3+}$, $Sc^{3+}$, $Ga^{3+}$, $In^{3+}$, $La^{3+}$ | $Tl^{3+}$, $Tl(CH_3)_3$, $BH_3$, $Ga(CH_3)_3$ |
| $N^{3+}$, $Cl^{3+}$, $Gd^{3+}$, $Lu^{3+}$ | $GaCl_3$, $GaI_3$, $InCl_3$ |
| $Cr^{3+}$, $Co^{3+}$, $Fe^{3+}$, $As^{3+}$, $CH_3Sn^{3+}$ | $RS^+$, $RSe^+$, $RTe^+$ |
| $Si^{4+}$, $Ti^{4+}$, $Zr^{4+}$, $Th^{4+}$, $U^{4+}$ | $I^+$, $Br^+$, $HO^+$, $RO^+$ |
| $Pu^{4+}$, $Ce^{4+}$, $Hf^{4+}$, $WO^{4+}$, $Sn^{4+}$ | |
| $UO_2^{2+}$, $(CH_3)_2Sn^{2+}$, $VO^{2+}$, $MoO^{3+}$ | $I_2$, $Br_2$, $ICN$, etc. |
| $BeMe_2$, $BF_3$, $B(OR)_3$ | trinitrobenzene, etc. |
| $Al(CH_3)_3$, $AlCl_3$, $AlH_3$ | chloranil, quinones, etc. |
| $RPO_2^+$, $ROPO_2^+$ | tetracyanoethylene, etc. |
| $RSO_2^+$, $ROSO_2^+$, $SO_3$ | O, Cl, Br, I, N, $RO^\bullet$, $RO_2^\bullet$ |
| $I^{7+}$, $I^{5+}$, $Cl^{7+}$, $Cr^{6+}$ | $M^\circ$ (metal atoms) |
| $RCO^+$, $CO_2$, $NC^+$ | bulk metals |
| HX (hydrogen bonding molecules) | $CH_2$, carbenes |

Borderline

$Fe^{2+}$, $Co^{2+}$, $Ni^{2+}$, $Cu^{2+}$, $Zn^{2+}$, $Pb^{2+}$, $Sn^{2+}$, $Sb^{2+}$, $Bi^{3+}$, $Rh^{3+}$, $Ir^{3+}$, $B(CH_3)_3$, $SO_2$, $NO^+$, $Ru^{2+}$, $Os^{2+}$, $R_3C^+$, $C_6H_5^+$, $GaH_3$

TABLE 4.3.   HSAB classification of bases (from PEARSON, R. G., *J. chem. Educ.*, **45**, 581, 643 (1963). The symbol R stands for an alkyl group such as $CH_3$ or $C_2H_5$

| Hard | Soft | - |
|------|------|---|
| $H_2O$, $OH^-$, $F^-$ | $R_2S$, $RSH$, $RS^-$ | |
| $CH_3CO_2^-$, $PO_4^{3-}$, $SO_4^{2-}$ | $I^-$, $SCN^-$, $S_2O_3^{2-}$ | |
| $Cl^-$, $CO_3^{2-}$, $ClO_4^-$, $NO_3^-$ | $R_3P$, $R_3As$, $(RO)_3P$ | |
| $ROH$, $RO^-$, $R_2O$ | $CN^-$, $RNC$, $CO$ | |
| $NH_3$, $RNH_2$, $N_2H_4$ | $C_2H_4$, $C_6H_6$ | |
| | $H^-$, $R^-$ | |

Borderline

$C_6H_5NH_2$, $C_5H_5N$, $N_3^-$, $Br^-$, $NO_2^-$, $SO_3^{2-}$, $N_2$

suggested further reading. One of the first observations of the HSAB principle at work was made by Berzelius who noticed that some metals occur on the earth's surface as ores of carbonate or oxide whereas other metals occur as sulphides. The explanation is that hard acids, e.g. $Mg^{2+}$, $Al^{3+}$, $Ca^{2+}$, form strong bonds with hard bases, e.g. $O^=$ or $CO_3^=$. Also, the softer acids, e.g. $Cu^{+/2+}$, $Hg_2^{2+}$ or $Hg^{2+}$ or $Pb^{2+}$ prefer soft bases, e.g. $S^=$. Hard acid–soft base or soft acid–hard base combinations, according to HSAB, will not have strong bonds and so their ores will have hydrolysed away millions of years ago.

Another use of HSAB is in helping us to remember and predict a large number of apparently unrelated phenomena: the hardness of an element obviously increases with oxidation state. Thus to stabilize an element, metal or non-metal, in a high oxidation state, it ought to be surrounded by hard bases such as oxide, hydroxide or fluoride. To stabilize an element in a low oxidation state it ought to be coordinated to soft bases such as carbon monoxide, phosphines, cyanides and arsines. This has obvious implications for the metal ions undergoing redox reaction *in vivo*. At first sight it may have been disturbing that many of the metal ions found *in vivo* are listed in Table 4.2 as borderline; however, because of the phenomenon of symbiosis this is actually an asset because it means that the metals can be called hard or soft depending upon whether their environment is one of hard or soft bases. *Symbiosis* is the process whereby one hard base on a metal ion encourages other hard bases to join it. Similarly, soft bases tend to flock together on a given acid centre also. An example of symbiosis may be seen in cobaltic complexes: $[Co(NH_3)_5X]^{2+}$ has stronger Co—X bonds for $X = F^-$ than for $X = I^-$. Hence, in this complex with hard ammonias, Co(III) is a hard acid. However, $[Co(CN)_5X]^{3-}$ is more stable when $X = I^-$ or $H^-$ than when $X = F^-$. Therefore the soft cyanide complexing ligands have symbiotically made Co(III) into a soft acid. In fact this symbiosis is so advanced that the complex with $X = F^-$ is not known. A biological example of symbiosis occurs in the zinc complex of carbonic anhydrase. This metalloenzyme binds halide anions $I^- > Br^- > Cl^- > F^-$ and so the zinc is evidently soft. In aqueous solution, however, the hydrated zinc ion binds the ions $F^- > Cl^- > Br^- > I^-$. Clearly, in water the hard solvation sphere symbiotically renders the zinc harder whereas in the enzyme environment it is made softer.

We shall see later that metal ions in biological systems are often in a state of suspension between two different oxidation states. The lower state can be stabilized symbiotically by adding soft ligands and the higher oxidation state by adding hard ligands. However, if we add *very soft* or *very hard* ligands the metal ion will be completely anchored in one oxidation state and so the living process (e.g. a redox reaction) is prevented. This, of course, is called poisoning and the poisons that are most well known are usually acids or bases that are so strongly held to the active sites of an enzyme that the sites are effectively blocked off. Examples of soft acid poisons are organic mercurials and cadmium ions. As well as site blocking, poisoning by heavy metal ions usually results in precipitating the metal–protein complex as well. These soft acids bind strongly to sulphur groups and so rob the organism of sulphur containing proteins such as those involving cysteine residues. Very soft base poisons function by robbing our bodies of metal ions. Examples include cyanides, sulphides, trivalent arsenic compounds and carbon monoxide. In small concentrations these poison by blocking, by becoming attached to the metals in metalloporphyrins and copper enzymes. At higher concentrations they remove the metal ion entirely from the enzyme.

Conversely, very non-poisonous or inert metals are needed when artificial pieces have to be introduced during surgery and so the pure metal which, if dissolved, would give soft ions are chosen, e.g. gold, silver, tantalum and platinum. Because they would give soft ions there is negligible tendency for these metals to give up electrons to form *soft* metal ion–*hard* solvent bonds ($H_2O$ is hard).

4.1.4 *Defining the magnitude of hardness and softness* If we could put all acids and bases in order of merit, right through from very hard to very soft, we would be in a position to predict the *relative* strengths of any combination of acids and bases. It ought to be noted that it is the relative rather than absolute strengths that really interest us as complex formation in solution involves replacing one ligand (often water) by another, and medication is the chemical replacement of an undesired reaction by a more tolerable one. Further, the bonds formed by species at the extremities of softness would be purely covalent ($S_A = S_B = 0$), and of hardness would be purely ionic ($\sigma_A = \sigma_B = 0$).

This hardness–softness order can be established by studying the enthalpies and entropies of reactions in solution (viz. Table 4.4):

TABLE 4.4. Heats and entropies of formation in aqueous solution correlated with HSAB classification

| Compound | $\Delta H^0$ kcal mol$^{-1}$ | kJ mol$^{-1}$ | $\Delta S^0$ kcal K$^{-1}$ mol$^{-1}$ | kJ K$^{-1}$ mol$^{-1}$ | HSAB classification |
|---|---|---|---|---|---|
| H$^+$  F$^-$ | 2·93 | 12·2 | 23·1 | 96·6 | hard–hard |
| H$^+$  CN$^-$ | – 10·4 | – 43·5 | 7 | 29 | hard–soft |
| Hg$^{2+}$ (CN$^-$)$_4$ | – 59·5 | – 248·9 | – 13 | – 54 | soft–soft |
| Hg$^{2+}$  I$^-$ | – 18·0 | – 75·3 | – 2 | – 8 | soft–soft |

$- RT \ln K = 2·303 \ RT. \ (S_A.S_B + \sigma_A.\sigma_B) = \Delta G^0 = \Delta H^0 - T\Delta S^0$. Both A and B are considered to be solvated. Hard–hard reactions are usually endothermic but entropy stabilized, i.e. many solvent-A and solvent-B bonds need to be broken before the reaction can occur and the energy to do this is greater than any energy liberated in forming the A—B bond. However, since there are many solvent molecules liberated the $\Delta S^0$ term is large. Soft acids or bases are only weakly solvated, if at all, and so soft–soft reactions involve very little energy to desolvate the reactants. Hence they are exothermic and have small or negative entropies (because two particles A + B are making one A—B).

Hence, in a strongly polar solvent such as water, the enthalpy change upon forming an acid–base bond will be more negative the softer the acceptor and donor involved. Thus, measuring the enthalpies of a standard acid with a range of bases gives the softness order for these bases and similar measurements on a range of acids with a standard base gives the order for the acids. Enthalpies and entropies of reaction and the methods of measuring these are discussed in a later chapter.

It would be convenient to list the softness order of all acids (or bases) in this book but, in practice, finding a standard acid (or base) that complexes with them *all* is not possible—for HSAB reasons! Further, if several bases are used as standards there are overlap difficulties in the table. However, when any multibase system is being considered it is usually not difficult to find a standard acid to suit all the bases under current discussion, e.g. Cobalt (III) in vitamin B$_{12}$ coenzyme is held by three different kinds of basic groups. In this case we merely let Co(III) be our standard acid and refer to tables of $\Delta H^0$ or log $K$ to see the softness order of these three bases. Many such tables appear in the recommended further reading. When one is considering one metal ion versus another, no matter which ligand is chosen as a reference, the softness order lies in the extended Irving–Williams series. This states that the binding order for divalent metal ions is Ca$<$Mg$<$Mn$<$Fe$<$Co$\ll$Zn$\ll$Ni$<$Cu and clearly this ranges from hard acid Ca$^{2+}$ to borderline Cu$^{2+}$. It is important to notice that the increments in stability from one metal ion to its neighbour depend upon the softness of the basic groups attached to them, e.g. pyrophosphate groups (P$_2$O$_7^{4-}$) are poor donors and show little change in metal ion from one metal to the next. On the other hand, enol groups

are good donors and exhibit a wide range of stabilities from Ca$^{2+}$ through to Cu$^{2+}$. Consequently as we change protein donor ligand atoms from all oxide (hard) to all sulphide (soft) we traverse the series of metal ions (see Table 4.5).

TABLE 4.5. The relationship between the protein donor atom and the metal it prefers (from reference [2])

| Donor atoms | —O$^-$ | $\equiv$N, —O$^-$ | $\equiv$N, —S$^-$ | —S$^-$ |
|---|---|---|---|---|
| Metals preferred by the protein | Ca, Mg, Mn | Fe, Co, Ni | Cu, Zn | Cd, Pb, Hg |
| Examples of enzymes | ATP-ases, Enolases | Carbonican-hydrase, oxyhemo-globin | Carboxy-peptidiase | So strongly bound that they poison enzymes |

Hence HSAB goes a long way towards answering question 1 on p. 41, in that it does predict which metals from Chapter 2 will be found associated with which ligands from Chapter 3, e.g.

1. Na$^+$, K$^+$, Ca$^{2+}$, Mg$^{2+}$: These are hard acids and prefer hard bases such as water (they are all strongly solvated in solution),

oxygen and nitrogen donor atoms such as one finds in the hydroxy groups in cell walls, carbonates, phosphates, glutamates, oxalates, lactates or in amines such as calcitonin (see later).

2. Transition metal ions: When redox reactions occur ($Fe^{2+/3+}$, $Cu^{+/2+}$, $Mn^{2+/3+}$ or $Co^{2+/3+}$) the higher oxidation state is stabilized by a harder environment than is the lower. In fact, the $B_{12}$ just mentioned has $Co^{3+}$ stabilized by at least four hard nitrogen donors. Softer ions such as $Zn^{2+}$ or Mo(II) are complexed to softer ligands such as the $RS^-$ groups in alcohol dehydrogenase or xanthine oxidase.

Finally, intelligently guessing bond strengths from HSAB is only second best to actually measuring them in the laboratory. Unfortunately, as the systems investigated become more complex and exciting, their bond strengths become experimentally more evasive to measure. Thus the quantitative approaches to HSAB appear to have a promising future.

### 4.2 *Bonds between metal ions, amino acids, peptides and proteins* [3]

It is convenient to review here our present knowledge of metal ion–amino acid (or polymer) interactions as these frequently exhibit HSAB principles.

4.2.1 *Amino acid complexes*    $NH_2$—CHR—COO$^-$ has three donor groups, N: O$^-$ and $=$O, but this last keto group rarely complexes to the same metal ion as the first two groups as this involves a four membered chelate ring, unless it is a hemichelate structure, involving water, is formed, viz.

Hence in general, amino acid complexes either involve a metal carboxylate salt or an amine complex or both, in which case we have a five membered chelate ring.

Co, Ni, Cu, and Zn(II) prefer to form chelate rings whereas the harder acids Mg and Ca(II) form salts with the carboxylate grouping.

4.2.2 *Simple peptides* Simple peptides (amino acids where the amine group is substituted with another amino acid radical) combine less strongly with metal ions than do their constituent simple amino acids, e.g. $\log \beta$, for Cu(II) glycylglycinate $= 6\cdot0$ whereas for Cu(II) glycine $= 8\cdot6$. This possibly arises because changing the proton affects the inherent strength factor (the basicity) of the $\alpha$ amine nitrogen atoms. Complexes in which some coordination positions are still aquated can undergo hydrolysis and then oxidation by molecular oxygen, e.g.

$$Co\,(glycylglycinate)_2(H_2O)_2 \rightleftharpoons Co\,(GG)_2 \cdot OH^- \cdot H_2O + H^+$$

$$2Co\,(GG)_2 \cdot OH^- \cdot H_2O + O_2 \rightleftharpoons \left[ (GG)_2 - Co \underset{O-O}{\overset{OH \cdots O^H}{<>}} Co - (GG)_2 \right]^{2-}$$

4.2.3. *Amino acids with reactive side-groups* If these side-groups are positively charged, e.g. as in arginate, the complexes are much less stable because of repulsive effects. If the side-group is negatively charged, e.g. as in glutamate, the extra charge can be used for dimerization to give a strong complex:

(Note the bridging water group)

The amino acids histidine and cysteine/ine have reactive side groups and so are the strongest complexing amino acids found it blood. Histidine can complex in four different ways (see Fig. 4.1), giving us the choice of a five, six, or seven membered ring. However, the situation does rationalize itself in that some metals prefer

N to O donors, some metal geometries cannot accomodate the larger rings and also the competing proton concentration (i.e. pH) dictates and narrows the range of possible complexes. There are many ways of investigating these systems, the results of an infra-red investigation [4] are shown in the figure and the method is

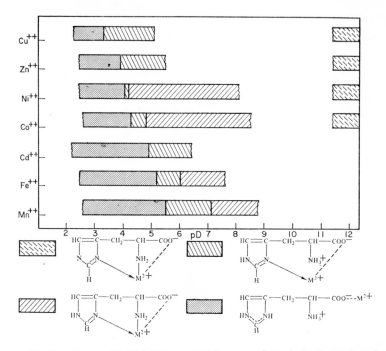

FIG. 4.1. Structures of 1:1 histidine–metal ion complexes in $D_2O$ as a function of pD. Open regions in the mid-pD range are indicative of low solubility. Taken from CARLSON, R. H. and BROWN, T. L., *Inorg. Chem.*, **5**, 268 (1966). (Copyright by the American Chemical Society. Reprinted by permission of the copyright owner.)

described in Chapter 8. Cysteine has a soft sulphur as a side-group and combines with soft acceptors more strongly than does any other amino acid. A redox reaction with cupric ions easily oxidizes cysteine to cystine

$$4RS^- + 2Cu^{2+} \rightleftharpoons 2RSCu + RSSR$$

Other complexes involving the reactive side-group include

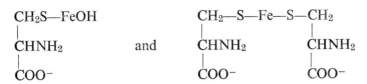

4.2.4 *Protein–metal ion interactions* There are seven factors to consider.

1. The chains are longer than in simple peptides so side-chain groupings ($-COO^-$, imidazole, $-SH$) are more important because their numbers greatly exceed those of terminal groups ($-NH_2$ and $-COO^-$).

2. There are always more binding groups than are actually needed to hold each metal ion so (a) it may choose which sites it prefers (on a HSAB basis) and (b) a statistical approach ought to be applied to the equilibria.

3. The tertiary structure of the protein prevents the chain from twisting and turning so proteins complex less readily than with simple model peptides. Hence, a protein having very many ligand groups may have only one or two metal ions attached to them and occasionally a metal ion may be held by only one group.

4. The protein contains positive and negative charges (e.g. $-NH_3^+$ and $-S^-$) that sometimes hinder and sometimes help the approaching metal ion, even though they are not directly involved in its eventual bonding.

5. Some side-groups are so arranged as to make the binding of a metal ion sterically very easy, i.e. every group is in just the correct position. These groups can collectively be called an active centre. At other times the donors are not quite so well arranged and many endeavour to hold the metal ion between two different coordinate bond geometries.

6. Any group that can donate electrons to a metal ion can also do so to a proton so that competition between metal ions and protons is important.

7. Bond formation involves desolvation of the site and of the metal ion.

Scatchard has examined all these factors in terms of probabilities

E

and has derived equations of binding. The subject is rather advanced and is described well in reference [3].

## REFERENCES

[1] PEARSON, R. G., *J. chem. Educ.*, **45**, 581 and 643 (1963).
[2] VALLEE, B. L. and WILLIAMS, R. J. P., *Chemy Brit.*, 397 (1963).
[3] GURD, F. R. N. and WILCOX, P. E., *Adv. Protein Chem.*, **2**, 311.
[4] CARLSON, R. H. and BROWN, T. L., *Inorg. Chem.*, **5**, 268 (1966).

## FURTHER READING

PEARSON, R. G. and DRAGO, R. S., *Chemy Brit.*, **3**, 103 and 516 (1967).
HUDSON, R. F., *Co-ord. Chem. Rev.*, **1**, 89 (1966).
BASOLO, F. and PEARSON, R. G., *Mechanisms of Inorganic Reactions*, 2nd edn (Wiley, New York, 1967).
AHRLAND, S., *Structure and Bonding*, **1**, 207 (1966); **5**, 118 (1968).
AHRLAND, S., *Helv. chim. Acta*, **50**, 306 (1967).

*Chapter 5*

# REMOVING UNWANTED METAL IONS

---

ALTHOUGH SODIUM, potassium, magnesium, and calcium together constitute 99·5% of the total metal content of our bodies, it is, in fact, the other metals present that are most likely to poison us because much smaller quantities are needed to double their concentrations. A discussion of the methods of removing these excesses is inherently related to the mechanisms of their toxic reactions since any detoxifying sequestering reagent must compete with these mechanisms. Table 5.1 lists the toxicities and poisoning symptoms of many metal ions. Our environment is polluted so we have many more metals present than are actually required for our metabolism, e.g. coastal dwellers will probably contain traces of most of the 54 elements found in sea-water. Nature has provided her own methods of protection against metal ion poisoning. First, except for sodium, potassium and calcium, metal ions are not readily absorbed by the gastrointestinal tract. Thus ingestion does not necessarily lead to poisoning. Secondly, unrequired metals are slowly filtered from the blood stream by kidney action. It is important to realize that even though an organism is dependent upon a metal ion for a metabolic process, it usually is not tolerant to excesses, e.g. extracellular potassium $= 4$–5 mM, 8 mM causes nerve and muscular abnormalities resulting in cardiac depression and eventual death. (Compare this tolerance with those of lead and aluminium listed in the table shown in Table 5.1 [1].)

The mechanisms of the toxicity may be divided into two broad regions.

TABLE 5.1. Mode of toxicity and sequestering agents for a variety of metal poisons

| Metal | Toxicity and physiological effect | Sequestering agent |
| --- | --- | --- |
| Na(I), K(I), Ca(II) | Increased osmotic pressures. Easily absorbed by intestine. Excess $Na^+$ causes Chinese restaurant syndrome (p. 26). | $Ca^{2+}$ may be removed by aminopoly-carboxylic acids such as EDTA. |
| Be(II) | Skin lesions and pulmonary damage. Be prefers O to N and S donors and so attacks phosphatase enzymes. Lethal dose = 1 part in $10^6$ of body weight. | Aurintricarboxylic acid. |
| Fe(II) and Fe(III) | Excess give siderosis effects (see Chapter 2). | $Na_2[Ca\ EDTA]$. |
| Co(II) | (i) inhibits —SH enzymes and (ii) causes polycythemia (increase in number of red blood cells). | (i) cysteine. |
| Cu(II) | Excesses arise from Wilson's disease (Cu control mechanism deranged) and Cu is deposited in liver and other parts. | Penicillamine or $Na_2[Ca\ EDTA]$. |
| Zn(II) | also<br>gives emesis and gastrointestinal irritation. | Diphenylthiocarbazone but exesses are toxic and also destroy the Zn–cysteine complex on the surface of the eyes. |
| Pb(II) | Normal blood conc. 1·5 µM.<br>Toxic blood conc. 4·0 µM } compare toxicities | BAL or $Na_2[Ca\ EDTA]$. |
| Al(III) | Normal blood conc. 18 µM<br>Toxic blood conc. 90 mM | |
| As(III), Hg(II), Cd(II), Au(I and III), Pb(II), Bi (III), Sb(III), V(II–V). | Attack —SH groups of essential enzymes. Hg(II) complexes are neutral, e.g. $HgCl_2$, so are very readily absorbed through intestine and so are very toxic. | BAL but the BAL Cd complex is toxic and causes renal damage. |
| U(IV) or (VI) | Poisoning by skin absorption. It becomes bound to bone but eventually it is carried to the kidney as bicarbonate complexes and then deposited. It kills the kidney cells as it is toxic to their enzymes. | $Na_2[Ca\ EDTA]$. |

1. $Li^+$, $Na^+$, $K^+$, $Ba^{2+}$ poison by electrolyte disturbance, absorption on negative sites of enzymes and osmotic imbalance.
2. Other metal ions poison by complex formation that obeys the HSAB rules, e.g. $Hg^{2+}$ searches out sulphur and nitrogen ligands, displacing weaker complexing metal ions if necessary. In fact, metal ion toxicity is related to the strength of the complex bond so that toxicity follows the Irving–Williams series in many instances, e.g. divalent metal ions upsetting urease and diastase activity.

Some general points are worthy of note concerning the sequestering agents in Table 5.1. Ethylenediaminetetraacetic acid (EDTA) is not the ideal reagent as (a) its rate of absorption from the intestine is slow so it is usually administered by intravenous injection as $Na_2[Ca\,EDTA]$ (the pure sodium salt is not used as it robs the body of essential calcium) and (b) its size hinders its progress towards the enzyme and also, being a very hydrophilic reagent, it does not relish going near enzymes anyway. Instead it complexes with the circulating metal ions and the ensuing circulatory deficiency of metals is replenished by removing them from the enzyme. Hence EDTA treatment is often a protracted process. 2,3-dimercaptopropanol (Fig. 5.1) (BAL), was originally developed

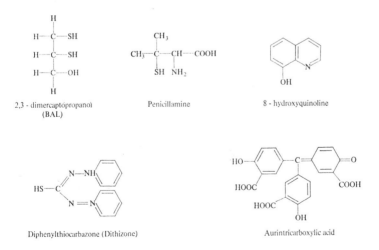

FIG. 5.1. Sequestering reagents used in removing metal ions from blood.

to protect —SH enzymes from war gas arsenicals. It also protects these enzymes against the metal ions listed against it in Table 5.1. It is ineffective against selenium poisoning as this element oxidizes the —SH groups rather than complexes with them. Sometimes sequestering is necessary to remove metal ions intentionally added, e.g. (a) [Pb EDTA]$^{2-}$ has been used as an X-ray contrasting agent and this dissociates to give Pb$^{2+}$ ions which are poisonous. (b) Gold complexes (e.g. sodium aurothiosulphate) are used to treat arthritis.

## 5.1 *Designing a complexing ligand for removing unwanted metal ions* [2]

Approximately 75% of all the elements are metals so it is not surprising that we should find metallic impurities inside our bodies from time to time and, in this respect, it is always much easier to get impurities into our bodies than to remove them. All the possible facets of solution chemistry must sometimes be invoked to rid ourselves of unnecessary ions. This is especially true now that more and more metal ions are being used in exotic commercial chemicals and as industrial catalysts. Although methods for purifying food are being improved, the range of metal ions that we are sometimes called upon to remove from the body is widening. Complexing molecules have to be specifically tailored so that they remove unwanted metal ions and leave the essential metals untouched. This tailoring is best achieved upon the basis of fundamentals so that the screening of thousands of possible sequestering reagents is reduced.

The principles of designing such a reagent are:
1. The ligand must be small enough to get through to the metal ion on the enzyme and form a strong enough complex to pull the metal ion away.
2. The reagent ought not to be destroyed by the method of introduction, e.g. if taken orally it must be robust enough to resist the destructive chemicals found in the digestive system.
3. Depending upon which metal ion we wish to remove, the donor groups are chosen on the basis of HSAB.
4. More powerful complex formation can sometimes be achieved if a ligand is used that has two or more donor groups that can

chelate (i.e. form a ring). Five and six membered chelate rings are usually the strongest and a ligand that can form two or more rings will give an even more stable complex.

5. The steric aspects of the metal ion must also be considered. For example, it is no use trying to remove Ag(I) by forming a five membered chelate ring because Ag(I) prefers two linear bonds and these are best achieved by eight membered rings or two monodentate ligands. Metal ions having tetrahedral or octahedral bonds prefer five or six membered rings. On the other hand, the sequestering ligand must not be so bulky that it cannot easily fit around a metal ion to form a chelate ring. Hence, experiments with atomic models ought to be tried before laboratory preparations are commenced. There is often a fine balance between the usefulness of side-groups for pushing electrons in the correct direction, but, at the same time, because of their size, these groups give steric hindrance problems, e.g. tetrahedral Cu(I) bis 2,9-dimethyl-1,10-phenanthroline is more stable than Cu(I) bis 1,10-phenanthroline. Hence, one might assume that tris Fe(II) 2,9-dimethyl-1,10-phenanthroline ought to be more stable than tris Fe(II) 1,10-phenanthroline but, in fact, it cannot be formed for steric reasons.

Sometimes a ligand can have more than two bonds to a metal ion, e.g. porphyrins (Fig. 2.4), and then the size of the central hole is very important. It is little use placing all the juiciest groups around a cavity to trap a trespassing metal ion if the ion is too large to fit into the cavity. Once again, atomic models can indicate whether the hole is too small or too large. Finally, it ought not to be forgotten that bond stereochemistries often differ between different valence states of the same metal ion. If a suitable sequestering ligand cannot be found for an ion, perhaps one is available for its reduced or oxidized form.

6. In laboratory experiments on living organisms, it is sometimes useful to see if the metal ion has indeed taken the bait and become attached to the ligand. Some ligand donor groups are chromophores (they give colour to a complex). These are indicated in Table 5.2. All other factors being equal, it is sometimes advantageous to choose ligands bearing these chromophoric groups.

7. If a ligand or its complex is to pass through a cell membrane

TABLE 5.2 Common donor groups found in metal ion sequestering reagents. Chromophores are indicated with an asterisk

and, for the moment, we consider this as being a non-aqueous phase, the ligand and its complex ought to be so designed as to be soluble in this phase, i.e. have an overall zero or small charge.

8. Choice of the ligand donor groups. The range of donor groups available in well-known sequestering reagents are listed in Table 5.2. Under the correct conditions and with the right metal ion these are known to form strong bonds and the reactions go far enough to completion to be quantitative. The offending metal ion is first examined from the point of view of its size, polarizability, and valence state to determine its hardness or softness. Then a range of donor groups are tentatively chosen. Next the formation constants between the metal ion and these groups are extracted from the literature [3]. The corresponding formation constants for interfering metal ions are also considered. In general, formation constants of transition metal complexes of any given ligand form a regular series but the slope of this progression depends upon the HSAB nature of the ligand donor atom, e.g. with N donors, Mn(II) is much less stable than Cu(II) complexes but with O donors, Cu(II) has only a slightly larger formation constant than does Mn(II). Hence, to remove the copper and leave the manganese, N donor chelate forming ligands would be suggested from the groups available in Table 5.2. Looking into the future upon a HSAB basis, if phosphorus, arsenic or selenium could be used as soft base donor atoms, the formation constant

separation for the softer metal ion impurities would be large and so easier selectivity would occur. However, stable ligands of these donor atoms are not yet available.

9. The stoichiometric ratio of ligand to metal in the complex must be carefully selected, e.g. it is possible to form the mono, bis and tris (8-hydroxyquinoline) iron (III) complexes but it has been suggested that the mono and bis are toxic and cannot penetrate cells whereas the tris is non-toxic and can penetrate cells because it is uncharged [1]. Hence, exact doses are imperative and this is a good example of the uses of computerized medication.

10. Solubility. Clearly the ligand, and any complex with any of the metal ions it may encounter *in vivo*, must be soluble. Water is a polar solvent that is strongly hydrogen bonded and for a substance to be soluble in water it must be capable of breaking some of these hydrogen bonds. The donor atom order for hydrogen bond breaking follows electronegativity and is $O > N \gg S \gg C$ and this implies that, although ideally we require as many formal charges on the molecule as possible, it is possible to choose donor groups from the left-hand side of this series and still have very soluble ligands and complexes because of the $\delta -$ (and corresponding $\delta +$) induced by the highly electronegative oxygen and nitrogen donors. Solubility is associated with charges and pH can govern the amount of charge on a complex, e.g. consider copper bis(glycinyl) complexes

(The pH decreases from left to right)

Further, slight changes in pH can have profound effects upon the concentration of a particular complex present (see Fig. 5.2).

One ought not to confuse solubility with dissolving power, although both are interrelated, e.g. tyrosine contains an extra —OH group compared with phenylalanine (see Fig. 3.1) and yet the former is very difficult to dissolve. This is believed to occur because the —OH helps to form hydrogen bonded 'chains' in the crystal so uniting one tyrosine molecule to another

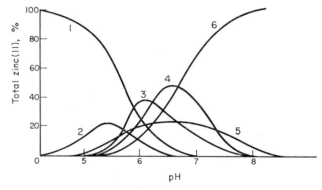

However, once sufficient energy has been provided to overcome this extra crystal lattice energy stabilization the —OH group that caused this can be used to hold the solid in solution. Hence complexes or ligands containing groups that have lattice building properties have the choice of forming a stable crystal lattice or of

FIG. 5.2. Variation with pH of the composition of solutions of cysteine (0·01 M) and zinc (II) ions (0·005 M). If $H = H^+$, $Zn = Zn^{2+}$ and

$$A = HS—CH_2—\underset{\underset{NH_2}{|}}{C}—COO^-$$

curves 1–6 show the percentage of total zinc present as free metal ion, $Zn(HA)_2$, $Zn_3A_3(HA)^-$, $Zn_3A_4^{2-}$, $ZnA(HA)^-$ and $ZnA_2^{2-}$, respectively. From Perrin and Sayce [4].

being in a well solvated dissolved state. In general the former is usually the case but the latter can be tailored by either invoking steric effects that upset crystal lattices or by adding strongly ionized groups to the molecule, e.g. sulphonic acid groupings.

Two general statements may be made concerning solubility. Complexes are usually less soluble than their ligands because if a lone pair of electrons is used to bind a metal ion, these electrons are no longer available to form hydrogen bonds to the solvent. When we are discussing the removal of precipitates *in vivo* (e.g. cataract), speed of treatment is sometimes important if the precipitate is to be eluted away. This arises when a precipitate can exist in two crystalline forms because usually the metastable form is the first to precipitate and the more soluble. In the absence of rapid treatment, the metastable form changes to the less soluble stable form. Further, the larger the crystal that is formed, the smaller the surface area per unit weight and therefore the slower the process of elution.

11. Dangers associated with metal ions and complexes *in vivo*. It is, of course, essential that the reagent chosen is low in toxicity and that its complexes are also non-poisonous, e.g. we wish to avoid using the zinc or cadmium complexes formed in the presence of excess 2,3-dimercaptopropanol because these are more poisonous than the free metal ions alone. However, these difficulties can usually be detected by efficient screening. Nevertheless, there are certain long term effects that screening cannot always detect. For example, the ligand or metal complex may be a carcinogen [5], [6]. Cancer is the autonomous growth of new abnormal body tissues and is reported to account for 20% of human deaths. Unfortunately, from the screening viewpoint, carcinogenesis is a slow process occurring over $0 \cdot 1 - 0 \cdot 5$ of the lifespan of an organism, so that if a person is exposed to a new ligand today, it may be five to forty years before cancer is evident. There is reason for postulating that metal ion complexation and cancer formation may be interrelated because many carcinogens have characteristic groups common to good donor ligands. Ideally we ought to list groups that are liable to cause cancer and then we can avoid using them in choosing a ligand (aromatic amines and nitroso compounds would probably head such a list). However,

(a) until a reasonable set of theories about the mechanism of cancer promotion and growth are available, this cannot safely be done;

(b) at the present time, it appears to be not so much the type of donor group, but rather the arrangement of groups that causes cancer (see Fig. 5.3); and

(c) other materials, called promoters, are sometimes required

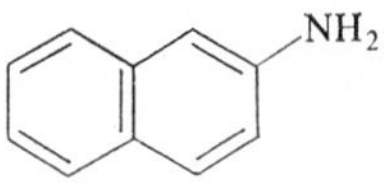

2 -Napthylamine 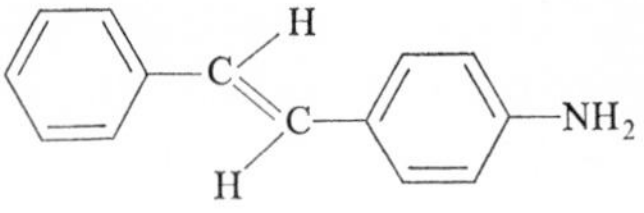                            4 Aminostilbene

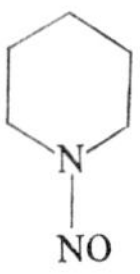

N − Nitrosopiperidine 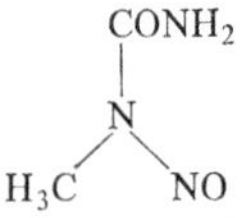              N-Methyl-N-nitrosourea

FIG. 5.3. Four known carcinogens. From Searle [5].

to persuade unsuspected groups to become carcinogenic, e.g. tobacco smoke is thought to contain promoters that catalyse latent lung cancer.

Questions facing future researchers into this deadly topic are as follows.

(a) Do carcinogens react with metal ions?

(b) Does the mechanism which changes a 'normal' chemical into a carcinogen involve metal ions?

(c) Is cancer treatment likely to seriously reduce the body's supply of metal ions?

(d) Is it possible to invoke metal ions to extend the kinetics of cancer to greater than fifty or sixty years or to cause a patient to react effectively against his own tumour?

## 5.2 *Medical uses of metal chelates in controlling virus infections* [1]

A virus may be described as a core of nucleic acid encased in a layer of protein, lipid or polysaccharide. It can only reproduce inside living cells where it takes over the chemicals and the energy supplies to the cell and uses them to produce more virus. Virus attacks occur via absorption of virus on to the cell, penetration of the cell and then the virial nucleic acids direct the synthesis of fresh virus (one virus can produce approximately one thousand others). These new viruses are then released from the cell and the host cell dies.

Inert, stable chelates have considerable anti-virus activity even against microorganisms that are resistant to antibiotics. Furthermore, most chelates are non-irritant at the therapeutic levels required and the organism usually cannot build up a resistance to these chelates. Hence, metal chelates are frequently used in controlling skin and wound infections. Their mode of action interrupts the process of virus invasion by (a) destroying the virus outside the cell, (b) occupying sites on the cell surface and so blocking virus entry, or (c) preventing the reproductive process inside the cell. In practice (c) involves killing the cell itself because we cannot chemically destroy the virus without harming the host cell.

The metal chemotherapy approach has so far been reasonably successful especially when the metal ion is administered as a complex of 1,10-phenanthroline. Commonly employed virus inactivators have been [Fe $(3,5,6,8$-tetramethyl-1,10-Phenanthroline$)_3]^{2+}$ and mixed complexes of the heavier element in the same group as iron, ruthenium (II), e.g. acetylacetone bis(3,5,6,8-tetramethyl-1,10-phenanthroline) Ru(II) chloride.

Once the complete set of mechanisms of virial infection are elucidated, the range of available sizes, charges, stereochemistries and redox potentials of metals and their ions can be exploited to provide cornerstones for tackling many microorganisms. Although knowledge itself cannot be directly related to the ability to cure, it certainly helps.

# REFERENCES

[1] *Chelating Agents and Metal Chelates.* Eds DWYER, F. P. and MELLOR, D. P. (Academic Press, London, 1964).

[2] PERRIN, D. D., *Organic Complexing Reagents* (Interscience, New York, 1964).

[3] SILLÉN, L. G. and MARTELL, A. E., *Stability Constants of Metal Ion Complexes.* Special Publication No. 17. 2nd edn (The Chemical Society, London, 1964) (3rd edn in preparation).

[4] PERRIN, D. D. and SAYCE, I. G., *J. chem. Soc.* (A), 53 (1968).

[5] SEARLE, C. E., *Chemy Brit.*, 5 (1970).

[6] STOCK, J. A., *Chemy. Brit.*, 11 (1970).

*Chapter 6*

# THE DETERMINATION AND USE OF FORMATION CONSTANTS†

A CONSIDERATION of formation constants must take into account (a) that water molecules, ligands (including OH⁻) and outer sphere complexing ligands are all competing for the metal ion, and (b) that the donor sites on the ligand attract water molecules and protons.

Thus, if we consider a complex formation reaction between a metal ion, B, and a protonated ligand, AH:‡

$$pAH(H_2O)_w + qB(H_2O)_x \rightleftharpoons A_pB_qH_r(H_2O)_y + (p-r)H(H_2O)_z + (w+x-y-z)H_2O \quad (6.1)\S$$

and the complex may then hydrolyse to give:

$$A_pB_qH_r(H_2O)_y \rightleftharpoons A_pB_qH_rOH(H_2O)_n + H(aq) \quad (6.2)\S$$

($n$ is not necessarily $y-1$ because replacing a $H_2O$ by a negative OH⁻ changes the overall charge on the complex and so it may be differently hydrated). The water in these equations is meant to be that in the inner spheres of hydration but pedantically the outer sphere ought to be considered also. In practice, the concentrations of free water $[H_2O]$ in aqueous solutions is assumed constant and is omitted from any calculations. However, it does act as an

---

† Formation constants and stability constants usually mean the same. In this review the term formation constants is adhered to, to stress that the constants refer to the *formation* of complex bonds.

‡ The A and B in this chapter do not have exactly the same meaning as the A and B in Chapter 4.

§ Charges omitted.

important buffer in reaction (6.1) and ought not to be forgotten when reactions not taking place in an excess of water are considered, e.g. such as inside cell membranes. The magnitude of this buffering by the free water in the bulk of the solution can be seen from a comparison of a simple formation constant in water and in one twentieth the concentration of water, e.g. in 95% methanol/water. Methanol only solvates approximately 10% as strongly as water and so 95% methanol/water has sufficient water present to adequately hydrate anything that needs hydrating. For the reaction

$$H^+(aq) + CH_3COO^-(aq) \overset{K}{\rightleftharpoons} CH_3COOH(aq) + H_2O \quad (6.3)$$

$K_{H_2O} = 3 \cdot 32 \times 10^4$; $K_{50\% \text{ dioxan}/H_2O} = 4 \cdot 15 \times 10^5$; $K_{95\% \text{ methanol}/H_2O}$
$$= 9 \cdot 23 \times 10^6 \text{ M}^{-1} \text{ at } 25°C \text{ and } 0 \cdot 500 \text{ M ClO}_4^-$$

Clearly decreasing the $[H_2O]$ in the bulk of the solution permits the reaction to proceed further to the right-hand side. Although we must bear this fact in mind, formation constants hereafter are quoted assuming $[H_2O] = 1$ throughout.

The formation constant for equation (6.1) is called $\beta_{pqr}$ and it applies to forming the products from the individual species, i.e.

$$pA + qB + rH \overset{\beta}{\underset{pqr}{\rightleftharpoons}} A_pB_qH_r \quad (6.4)$$

It is the product of the stepwise formation constants describing each stage in the formation of $A_pB_qH_r$ and because it is a constant it must be the same whichever stepwise scheme we use to define it, e.g.

| Scheme 1 | | Scheme 2 | |
|---|---|---|---|
| $A + B \rightleftharpoons AB$ | $K_1$ | $A + H \rightleftharpoons AH$ | $K_I$ |
| $AB + B \rightleftharpoons AB_2$ | $K_2$ | $AH + H \rightleftharpoons AH_2$ | $K_{II}$ |
| $\cdot$ | $\cdot$ | $\cdot$ | $\cdot$ |
| $\cdot$ | $\cdot$ | $\cdot$ | $\cdot$ |
| $AB_{q-1} + B \rightleftharpoons AB_q$ | $\cdot$ | $AH_{r-1} + H \rightleftharpoons AH_r$ | $\cdot$ |
| then $AB_q + A \rightleftharpoons A_2B_q$ | $\cdot$ | then $AH_r + B \rightleftharpoons ABH_r$ | $\cdot$ |
| $A_2B_q + A \rightleftharpoons A_3B_q$ | $\cdot$ | $ABH_r + B \rightleftharpoons AB_2H_r$ | $\cdot$ |
| $\cdot$ | $\cdot$ | $\cdot$ | $\cdot$ |
| $A_{p-1}B_q + A \rightleftharpoons A_pB_q$ | $\cdot$ | $AB_{q-1}H_r + B \rightleftharpoons AB_qH_r$ | $\cdot$ |
| then $A_pB_q + H \rightleftharpoons A_pB_qH$ | $\cdot$ | then $AB_qH_r + A \rightleftharpoons A_2B_qH_r$ | $\cdot$ |
| $A_pB_qH + H \rightleftharpoons A_pB_qH_2$ | $\cdot$ | $A_2B_qH_r + A \rightleftharpoons A_3B_qH_r$ | $\cdot$ |
| $\cdot$ | $\cdot$ | $\cdot$ | $\cdot$ |
| $A_pB_qH_{r-1} + H \rightleftharpoons A_pB_qH_r$ | $K_n$ | $A_{p-1}B_qH_r + A \rightleftharpoons A_pB_qH_r$ | $K_N$ |

$\beta_{pqr} = K_1 . K_2 \ldots K_n \equiv K_I . K_{II} \ldots K_N \equiv$ product of any similar scheme commencing with A, B and H and ending in $A_pB_qH_r$. If none of the $K$ values are infinity or zero, any solutions of A, B and H will contain traces of every complex mentioned in the above schemes. Fortunately many of the species are present in so low concentration that they may be ignored in any calculations.

Two problems now confront us: From solution chemistry investigations,

1. How do we decide which species are the main ones present?
2. How best can we express the formation of these major species in terms of formation constants?

Problem 1 may be answered by listing a series of generalizations that together add up to 'chemical intuition or experience'.

(a) HSAB, or similar theories, indicate whether strong or weak bonds between A and B are to be expected; if weak then B will not have many A attached to it, i.e. $p=0$ or low.

(b) If the range of complexes formed by one metal ion is known, we may use the extended Irving–Williams series to estimate the strength of complexes formed by another metal ion in the same series, i.e. the range of $p$ may be guessed. Listings of already determined constants have been published [1, 2] and these can often be a useful source for extrapolations.

(c) The configurations, and hence the maximum number of coordinate bond positions, of each of the metal ions have been recorded in Chapter 2, e.g. Co(II) has a maximum coordination number of six so $p$ cannot exceed six for mononuclear complexes. For polynuclear complexes ($q>1$), $p$ cannot exceed $6q$. An example of a polynuclear complex is shown in Fig. 6.1. If the ligand is

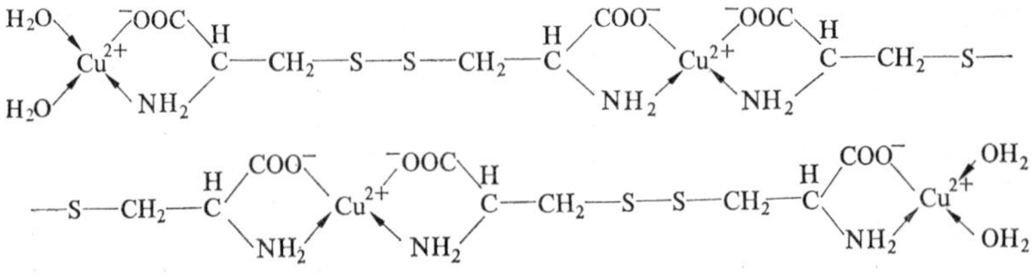

FIG. 6.1. An example of a polynuclear complex. The copper (II)–cystine polymer when $p=3$ and $q=4$.

bidentate, $p$ ought not to exceed $(6/2)q$. Similarly a maximum value for $r$ may be estimated by counting the number of donor positions on the ligand ($d$, say) and subtracting $2q$ from $dp$ (assuming bidentate ligand).

(d) Concentration dependence of the range of complexes present. The concentration of free water in equation (6.1) was seen to buffer the quantity of product formed. Similarly the amount of reactants pushing the reaction forward in equation (6.4) also helps to determine which species will be most important. The influence of this pushing power is illustrated in Fig. 6.2. Clearly the greater

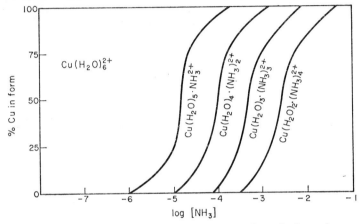

FIG. 6.2. The distribution of cuprammonium complexes is dependent upon the $NH_3$ pushing the reaction from left to right.

the concentration of $NH_3$, the larger the value of $p$. In the figure it changes from $p=0$, $Cu(H_2O)_6^{2+}$, to $p=4$, $Cu(NH_3)_4(H_2O)_2^{2+}$. Likewise, one would not expect to find protonated complexes ($r>0$) at high pHs or hydroxycomplexes at low pHs. Also if the total concentration of B, $T_B$, is greatly in excess of $T_A$, $p>q$ is unlikely and if $T_A \gg T_B$, $q$ values greater than $p$ may be assumed absent.

Clearly the composition of the complexes present will change with pH, e.g. O-phosphorylserylglycine–$Ca^{2+}$ and $Mg^{2+}$ complexes change their composition around the pH of blood from $p=q=r=1$ at pH$=6\cdot90$ to $p=q=1$, $r=0$ at pH$=7\cdot90$. The proton lost is the one shown with the positive charge. Further, if we then consider the copper (II) complex instead, Österberg has

F

$$\begin{array}{c} \text{CH}_2\text{—CH—C—NH—CH}_2\text{—COO}^- \\ \end{array}$$

shown that as we lose protons from the ligand (as pH increases), the $Cu^{2+}$ instead of being held mainly by Cu—O bonds changes to being held by Cu—N bonds [3]. These results are important when one considers why enzymes containing phosphorylserylglycine components are activated by calcium, magnesium and manganese (II) but inhibited by copper (II) which is bound $10^6$ times as strongly as the former.

Thus, allowing for a fairly large margin of error, we can list the possible complexes present and then use this list in answering question 2.

## 6.1 *Determination of formation constants* (problem 2)

Formation constants may be expressed in terms of activities, in which case they are known as thermodynamic constants, or concentrations, when they are called concentration constants, or sometimes in mixed activity and concentration terms (Table 6.1).

TABLE 6.1. Three different definitions of formation constants, often encountered in solution chemistry. The reaction is $A + H^+ \rightleftharpoons AH^+$

| Type of constant | Definition | Note |
|---|---|---|
| Concentration | $\beta_{HA} = \dfrac{[AH^+]}{[A][H^+]}$ | [ ] = concentration |
| Thermodynamic | $\beta_{HA} = \dfrac{[AH^+]}{[A][H^+]} \cdot \dfrac{f_{AH^+}}{f_A \; f_{H^+}}$ | $f$ = fugacity |
| Mixed | $\beta_{HA} = \dfrac{[AH^+]}{[A] \cdot a_{H^+}}$ | $a_{H^+} = f_{H^+}[H^+]$ = activity |

At zero ionic strengths ($I=0$) concentrations equal activities and all three definitions give thermodynamic constants. Unfortunately we can never investigate reactions in solutions of zero ionic strength as both reactants and products contribute to $I$. Hence concentration constants obtained at several different $I$ values have to be extrapolated to $I=0$ if we desire thermodynamic constants. This approach is not often used in practice as (a) it is not easy to fit the constants on to a smooth curve (see Fig. 6.3) [4], and (b)

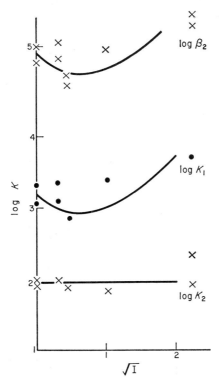

FIG. 6.3. Formation constants for the $Ag^+$–$Cl^-$ system as a function of ionic strength, $I$. Taken from Dyrssen, Jagner and Wengelin [4].

the experimental load in using this approach is often ten times (for ten $I$ values) the labour needed for determining constants at an agreed ionic strength. In fact trends found in constants at a fixed ionic strength are usually seen in thermodynamic constants also.

## 6.2 Choice of medium

If we are only dealing with neutral ligands the constants do not vary a great deal with ionic strength. In biological systems the ligands are not always neutral and so the problems are at what ionic strength to work and which ionic background salt to use.

In order to avoid the complications of ion pairing mentioned in Chapter 2, perchlorate or nitrate would be the best anions. The ionic mobilities of $K^+ \simeq NO_3^-$ whereas $Na^+$ has a much lower mobility. Thus potassium nitrate is a widely used background salt. For similar reasons potassium perchlorate would be advisable but unfortunately it is insoluble so the sodium or lithium perchlorates are used instead. The anion concentration is held constant rather than that of the cation. It was Brönsted in 1927 who first suggested using a high and constant concentration of an indifferent salt to hold the activities of reacting species constant and under these conditions one is justified in expressing the law of mass action in terms of concentrations. Dozens of other leading scientists have verified this statement since then.

Biedermann [5] has shown that when the sum of the equivalent concentrations of all the positive and negative ions disappearing during a complex forming reaction ($A^- + B^+ \rightleftharpoons AB$) does not exceed 0·15 M, the activity factors (fugacities) can be effectively held constant by 3·00 M $ClO_4^-$ as an ionic background medium. The Sillén school in Stockholm has reported hydrolysis constants at 25 °C for the majority of the metals in the periodic table in this medium. Their values are a valued asset to complex formation studies as they can immediately be incorporated into complex formation calculations. Constants determined in this 3 M $(Na)ClO_4$ probably exhibit the same trends in stabilities as those of the isotonic 0·15 M NaCl of blood but their individual values may not be the same. Hence, the Perrin school at Canberra uses 0·15 M $KNO_3$. Chloride is not used as it acts as a competing ligand. 0·15 M $NH_4NO_3$ has also been used because it has a theoretical advantage in that its structure least upsets that of water ($NH_4^+$ is isoelectronic with $H_3O^+$). However, $NH_4^+$ is the protonated form of yet another ligand that can compete for the metal ion. Hence for biological experiments the mediums just quoted are to be preferred. Many other media are quoted in the literature and there are many good reasons for using them. Nevertheless, whatever the medium, one criterion must be satisfied: the background salt used must be ultra-pure; 1 part in $10^5$ impurity of $B^{n+}$ in 3 M $NaClO_4$ when we are studying metal ion concentrations of 1 mM speaks for itself!

The medium having been chosen, we now approach the task of converting experimental results into formation constants. There are many methods of doing this and Rossotti and Rossotti [6] give an excellent review of the graphical methods. The very complicated ligands found in living systems and the powerful computers that have now been perfected have led to computational methods being widely used as a check and as an extension of graphical methods. Just two of the more important programmes available, LETAGROP [7] (Swedish for 'Least Squares'), and SCOGS [8] (Stability Constants of General Species), will be mentioned here.

### 6.3 Computing the constants

For any experimental measurement we have

(a) a set of accurately known values,
(b) a measured value or two, and
(c) a set of unknowns.

For example, consider a simple e.m.f. titration to determine the pK of an acid ($\beta_{HA}$). The e.m.f., $E$, may be measured on a glass electrode (giving [$H^+$]) or a ligand electrode (such as Ag/AgCl for following [$Cl^-$] if HA refers to HCl).† The e.m.f. for each point in the titration, $E^i$,

$$E^i = f(T, t, v^i \ v_0, T_A, E^0, \beta_{HA}, K_w) \qquad (6.5)$$

where $T$ is the temperature, $t$ is the concentration of titre, $v^i$ is the volume added, up to point $i$, to $v_0$ of initial solution, $T_A$ is the total ligand concentration, $E^0$ is the standard electrode potential for the electrode arrangement chosen, $\beta_{HA}$ is the required pK, and $K_w$ is the ionic product of the medium. Hence we can list our three groupings:

(a) $T, t, v^i, v_0, T_A, E^0, K_w$
(b) $E^i$
(c) $\beta_{HA}$

---

† Glass or Ag/AgCl electrodes really respond to $a_{H^+}$ and $a_{Cl^-}$ but it is usual when determining *concentration* constants to measure the $E^0$ values by taking electrode readings, not from $a_{H^+}$ buffers, but from solutions of known [$H^+$]. Remember, also, that $E^0$ is $I$ dependent so ought to be measured at one's chosen $I$.

This data is fed into a computer that has been instructed (=
programmed by LETAGROP and SCOGS) to make a series of guesses
at (c). For each guess, an $E_{calc}$ is computed for each of the $n$
points in the titration and the best $\beta_{HA}$ is chosen as being the one
that produces $U$, of the least squares sum,

$$U = \sum_{i=1}^{n} (E^i{}_{calc} - E^i)^2$$

to be a minimum, $U_{min}$.

If (c) consists of several unknowns, all are varied (guesswise) and
for *every* possible combination of guesses a value of $U$ is calculated so that, once again, $U_{min}$ can be chosen. For just one
unknown, the $U$ versus $\beta$ curve is as in Fig. 6.4. Initially, quite

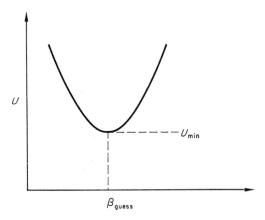

FIG. 6.4. $U$ versus $\beta_{guess}$ for a single unknown formation constant. Note that
the curve is a parabola.

large guesses at $\beta$ may be made and then when the region of
$U_{min}$ is found, the computer is instructed to search more closely,
using smaller jumps. For two unknowns, the $\beta_1$, $\beta_2$, $U$ curve
resembles the surface of a paraboloid (Fig. 6.5) around $U_{min}$.
Actually, many unknowns, $N$, say, can be handled by the computer but we would need $(N+1)$ dimensional space to discuss it
here. Both LETAGROP and SCOGS are capable of performing these
tasks and of printing out lists of the 'best' constants chosen by the
least squares technique. LETAGROP is a much larger programme

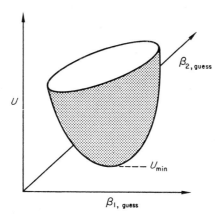

FIG. 6.5. $U$ versus $\beta_{1,\text{guess}}$ and $\beta_{2,\text{guess}}$ for two unknown formation constants.

than SCOGS and there are several points about LETAGROP that are worthy of mention.

1. The programme is completely general in its applications. Although originally written for finding the best set of formation constants, the procedure can actually be applied to any problem as long as the three groups (a), (b) and (c) and the function in equation (6.5) are known. The literature reports at least 26 fields to which LETAGROP has been applied. These range from n.m.r. studies to e.m.f. titrations on sea water.

2. In our simple example we placed the least squares error on $E^i$, the measured value and $v^i$ was placed in (a) as an accurately known value. This is the accepted procedure that the error square sum ought to be placed on the least accurately known parameter.

3. Once the 'best' set of constants have been found the $U_{\text{min}}$ can sometimes be decreased even further by assuming a systematic error in each titration and feeding this error in as an unknown constant, e.g. $t = 1 \cdot 037 \times \text{K}$ M NaOH, or pipette volume $v_0 = v_0 \times \text{K}$.

4. Having all the figures available, it is easy for the computer to calculate the standard deviation in the 'best constants' and in the titre.
   Then it can use these best constants to calculate the actual concentration of each species for each point in the titration.

5. LETAGROP can handle solid, liquid, and gaseous phases.

*Example* (*a*). Histidine hydroperchlorate monohydrate has a region of low solubility around physiological pHs (see Fig. 6.6)

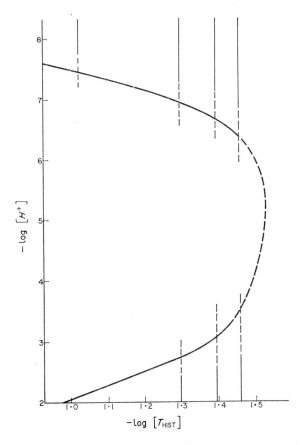

FIG. 6.6. Region of precipitation of histidine $HClO_4 \cdot H_2O$ shown on a $-\log[H^+]$ versus $-\log[T_{\text{Histidine}}]$ graph. The vertical lines indicate the pH range over which potentiometric titrations were possible. The broken line indicates the inconsistency of precipitation found between different measurements (this depends upon the rate of addition and stirring). The curved line is calculated from $\log K_S = -17 \cdot 65$ using HALTAFALL.

and this can be expressed in terms of a concentration solubility product [9] $\log K_S = [\text{Hist}^-] \cdot [H^+]^2 \cdot [ClO_4^-] = -17 \cdot 65 \pm 0 \cdot 13$ at $25 \cdot 0°C$, $3$ M $ClO_4^-$. To determine this value, $\beta_{HA}$, $\beta_{H_2A}$ were previously determined in solutions that were less than 30 mM in $T_{Hist^-}$ and then these values were fed in as (a) and $K_s$ was entered under (c) for titrations with $T_{Hist^-} > 30$ mM.

*Example* (*b*). When gases are involved, partial pressure constants may be used and solution formation constants and solubility products can also be incorporated into the same least squares computation. An example would include the decarboxylation of histidine (Fig. 6.7). The process primarily involves a

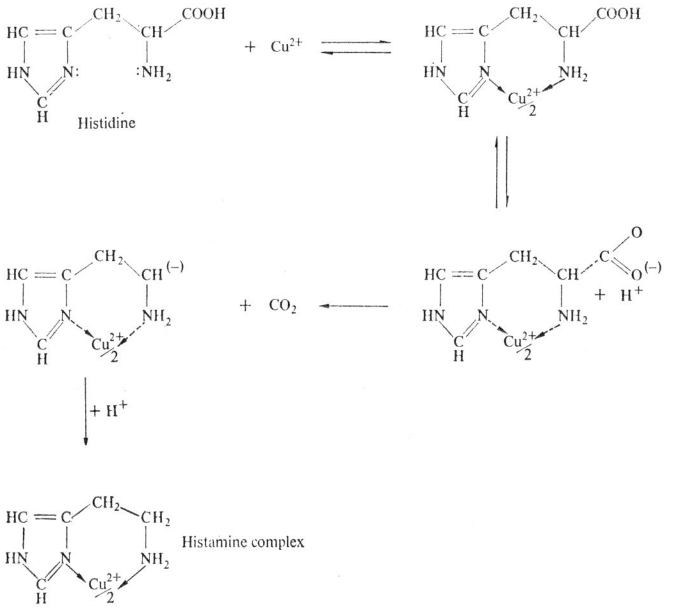

FIG. 6.7. The copper (II) catalysed decarboxylation of histidine to histamine and carbon dioxide. Taken from Andrews and Grundemeier [10].

chelation, the removal of —COOH as $H^+$ then $CO_2$ gas and finally the copper complex of histamine is formed. A LETAGROP investigation would indicate the reaction to be kinetically controlled, first order with respect to histidine, and the formation constants of the reactants are not the same as the products. In fact the histidine complex is much more stable than the histamine and so the latter when formed is always available to be broken up to give $Cu^{2+}$ ions to form a histidine complex again. This is why the $Cu^{2+}$ is known as the catalyst in this reaction [10].

### 6.4 *Uses of formation constants*

The uses may be summarized as follows:

1. To answer 'I wonder what would happen if . . .' queries.
2. To confirm a postulated mechanism.
3. As a precursor to enthalpy and entropy investigations—see Chapter 7.

1. When many reactions are occurring simultaneously in the same container, e.g. the blood stream, it is not possible to resolve them into their constituent formation constants *in situ* and then to calculate the effect of altering the system, e.g. changing the concentration of a metal ion or of administering a drug. Instead, we must resort to examining each possible metal–ligand reaction *in vitro* and characterizing all these reactions in terms of formation constants. Next we instruct a computer to combine all these reactions and to calculate the concentration of each complex at equilibrium. These concentrations ought to be checked against an analysis of the contents of the reaction vessel. Then the effect of adding our new reaction is calculated. Clearly this approach has a great future in computerized medication especially as many constants for the main reactions in blood are now known. The task resolves itself into determining the $H^+$ and $M^{n+}$ – drug formation constants and then computing the answer to 'I wonder what would happen if 1, 10, 20, 50 or 100 mg of drug were to be administered', i.e. how would the concentrations of all the complexes readjust themselves?

There are two widely known programmes for these calculations, COMICS [11] (Concentration of Metal Ions and Complexing Species) from Dr. Perrin's laboratory and HALTAFALL [12] (Swedish for 'Concentrations and Precipitates') from the late Professor Sillén's laboratory.

*Example.* Use COMICS to calculate the distribution of the copper (II) and zinc (II) complexes of the 16 amino acids, Glu(NH₂), Ala, Val, Gly, Pro, Leu, Thr, Ser, His, Ileu, ornithine (Orn), Try, Glu, CySSCy, CySH and Met. found in blood at pH = 7·4 at 37°C and $I = 0·15$ M.† The total concentrations of the

---

† Glu(NH₂) and Orn are derivatives of the carboxylic acid group in the side-chain of glutamic acid (see Fig. 3.1). The —COOH is changed to —CONH₂ for Glu(NH₂) and to —CH₂NH₂ for Orn.

amino acids vary from 0·57 mM Glu(NH₂) to 0·026 mM Met.
and the total concentrations of copper (II) = 0·018 mM and
zinc (II) = 0·046 mM [13].

The input data is shown in Fig. 6.8 and the computed output is
given in Fig. 6.9. It ought to be remembered that every formation

Plasma Composition (mM)

| | | | |
|---|---|---|---|
| GLU(NH₂) | 0·57 | HIS | 0·074 |
| ALA | 0·38 | ILEU | 0·068 |
| VAL | 0·25 | ORN | 0·055 |
| GLY | 0·21 | TRY | 0·054 |
| PRO | 0·21 | GLU | 0·048 |
| LEU | 0·13 | CYSSCY | 0·042 |
| THR | 0·12 | CYSH | 0·033 |
| SER | 0·11 | MET | 0·026 |
| COPPER (II) 0·018 | | ZINC (II) | 0·046 |

Then, there follows a list of the formation constants for the ticked (√) complexes:

Amino acid        Types of complexes

| Amino acid | CuA | CuA₂ | CuHA | CuHA₂ | ZnA | ZnA₂ | ZnA₃ | Zn(OH)A | ZnHA |
|---|---|---|---|---|---|---|---|---|---|
| ALA | √ | √ | √ | √ | √ | √ | √ | √ | |
| CYSH | | | √ | | | √ | | | |
| CYSSCY | | | √ | | | | | | |
| GLU | √ | √ | √ | | √ | √ | | | |
| GLU(NH₂) | √ | √ | | | √ | √ | | | |
| GLY | √ | √ | √ | √ | √ | √ | √ | √ | √ |
| HIS | √ | √ | √ | √ | √ | √ | | | √ |
| ILEU | √ | √ | | | √ | √ | √ | | |
| LEU | √ | √ | √ | √ | √ | √ | √ | | |
| MET | √ | √ | | | √ | √ | | | |
| ORN | √ | √ | √ | √ | √ | √ | | | √ |
| PRO | √ | √ | √ | √ | √ | √ | √ | √ | |
| SER | √ | √ | | | √ | √ | √ | | |
| THR | √ | √ | | | √ | √ | √ | | |
| TRY | √ | √ | | | √ | √ | √ | | |
| VAL | √ | √ | √ | √ | √ | √ | √ | √ | |

also ZnHA₂, Zn(HA)₂, Zn₃A₄ for CYSH; Cu₂A, Cu₂A₂ for CYSSCY; Cu(OH)A for ILEU; Cu(HA)₂, Zn(HA)₂, ZnHA₂ for ORN; Zn(OH)₂A for PRO;

and, finally, the formation constants for the following mixed complexes:

Cu(CYSH)(HIS)−           ZnH(CYSH)(HIS)
CuH₂(CYSSCY)(HIS)+    ZnH(CYSH)(HIS)₂−
CuH(CYSSCY)(HIS)      Zn(CYSH)(HIS)−
Cu(CYSSCY)(HIS)−     Zn(CYSH)(HIS)₂²−
Cu₂H(HIS)₂(CYSSCY)+   ZnH₂(CYSSCY)(HIS)+
                   ZnH(CYSSCY)(HIS)
                   Zn(CYSSCY)(HIS)−
                   Zn(GLU)(HIS)−
                   ZnH(GLY)(HIS)+
                   Zn(GLY)(HIS)

FIG. 6.8. The input information for the COMICS calculation of the complexes
present in a model of blood plasma. (Taken from Tables 1–3 of PERRIN, D. D.,
*Suomen Kemistilehti* 42(9), 205 (1969)).

## Copper

| Major species | % |
|---|---|
| CuH(HIS)(CYSSCY) | 45·4 |
| Cu(HIS)(CYSSCY)$^-$ | 39·6 |
| Cu(HIS)$_2$ | 10·7 |
| (Cu CYSSCY)$_2$ | 1·3 |
| Cu(GLU.NH$_2$)$_2$ | 0·65 |
| Cu(HIS)$^+$ | 0·31 |
| Cu(ALA)$_2$ | 0·25 |
| Cu(HIS)(CYSH)$^-$ | 0·25 |
| Cu(VAL)$_2$ | 0·22 |
| Total | 98·68 |

Free [Cu$^{2+}$] $1·1 \times 10^{-11}$ M
Free [Zn$^{2+}$] $5·0 \times 10^{-6}$ M

## Zinc

| Major species | % | Minor species | % |
|---|---|---|---|
| Zn(CYSH)$_2$$^{2-}$ | 20·6 | Zn(HIS)(CYSSCY)$^-$ | 0·95 |
| Zn(HIS)$^+$ | 19·3 | Zn(VAL)$^+$ | 0·88 |
| Zn(HIS)(CYSH)$^-$ | 14·0 | ZnH$_2$(HIS)(CYSSCY)$^+$ | 0·73 |
| Zn(HIS)(GLY) | 7·8 | Zn(LEU)$^+$ | 0·54 |
| Zn(GLU.NH$_2$)$^+$ | 4·1 | Zn(HIS)$_2$(CYSH)$^{2-}$ | 0·53 |
| Zn(HIS)$_2$ | 3·5 | Zn(GLU.NH$_2$)$_2$ | 0·39 |
| ZnH(HIS)(CYSSCY) | 3·2 | Zn(TRY)$^+$ | 0·38 |
| ZnH(HIS)(CYSH) | 2·1 | Zn(PRO)$^+$ | 0·35 |
| Zn(GLY)$^+$ | 1·8 | Zn OH$^+$ | 0·32 |
| Zn(THR)$^+$ | 1·6 | Zn(GLU) | 0·30 |
| ZnH(CYSH)$_2$$^-$ | 1·3 | Zn(HIS)(GLU)$^-$ | 0·30 |
| Zn(ALA)$^+$ | 1·2 | ZnH(ORN)$^{2+}$ | 0·29 |
| Zn(SER)$^+$ | 1·2 | Zn(ILEU)$^+$ | 0·23 |
| Zn$_{aq}$$^{2+}$ | 10·9 | ZnH(HIS)(GLY)$^+$ | 0·23 |
| Total | 92·6 | | |

Grand total 98·9%

Fig. 6.9. The computed output obtained from COMICS and the input data indicated in Fig. 6.8. (Taken from Table 4 of PERRIN, D. D., *Suomen Kemistilehti*, **42(9)**, 205 (1969)).

constant has been determined separately beforehand. The highest concentrations of complexes is seen to be CuH(His)(CySSCy) 45·4%, Cu(His)(CySSCy)⁻ 39·6%, Cu(His)₂ 10·7% and the free metal concentrations are $[Cu^{2+}] = 10^{-11}$ M and $[Zn^{2+}] = 10^{-6}$ M. A possible structure of the mixed complex present in highest concentration is illustrated in Fig. 6.10. It is interesting that this

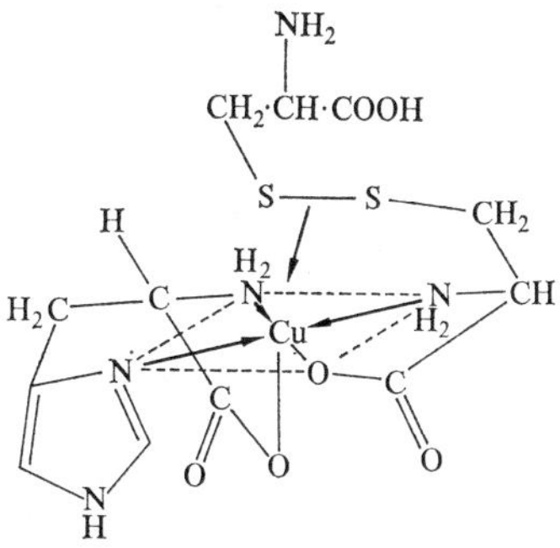

FIG. 6.10. Suggested structure for CuH(His) (CySSCy).

species is uncharged and so can be transported easily across membranes. It should be noted that this is an abbreviated example of the conditions in blood. A comprehensive study would involve all the ions of Chapter 2 and many of the ligands of Chapter 3.

Just like LETAGROP and SCOGS, COMICS and HALTAFALL calculate by continuously iteratively improving their guesses at the free metal ion and free ligand concentrations until both the equilibrium constants and the mass balance equations are satisfied to within a specified tolerance.

2. To illustrate the use of formation constants in confirming a postulated mechanism we shall use HALTAFALL, a very versatile programme that can consider solid, liquid and gaseous phases at one and the same time. As an example we shall take a system that is outwith, but nevertheless very relevant to, life in our bodies— the maintenance of the constant composition of the oceans and air [14, 15].†

† The constitution of the sea and the metals of life are not completely unrelated since our ancestors made the transition from aquatic to terrestrial life many thousands of years ago and hence blood and tissue fluids have similar contents to sea-water.

Textbooks have stated that the pH of sea-water ($8\cdot1 \pm 0\cdot2$) is determined by the $H_2CO_3 + HCO_3^- + CO_3^=$ reaction, the silicate sediments on the ocean floor and the atmosphere above being largely irrelevant to this pH stat process. In 1959, Professor Sillén suggested that the ocean, its floor, the land in between the oceans and the atmosphere above were all involved in one immense continuous titration; the oceans evaporating, coming down again as rain and the weather washing rocks and soil (both dissolved and suspended) into the sea. He further suggested that the result of this titration was that the composition of the sediment, solution and atmosphere remained approximately constant. 71% of the earth's crust is covered in water and the volume of the oceans is $1\cdot37 \times 10^{21}$ l. For the HALTAFALL calculations we use 1 l. of sea-water and *pro rata* to nature 3 l. of air ($0\cdot1010$ mole $N_2$, $0\cdot0271$ mole $O_2$, $0\cdot0012$ mole Ar and $0\cdot000039$ mole $CO_2$) and $0\cdot6$–$1\cdot2$ Kg sediments on the ocean floor. The system has the components listed in

TABLE 6.2. Balance of materials for the formation of one litre of sea-water, calculated from the estimates of Goldschmidt, V. M., *Fortschr. Miner. Kristallogr. Petrogr.*, **17**, 112 (1933), and Horn, M. K. and Adams, J. A. S., *Geochim. cosmochim. Acta.*, **30**, 279 (1966). Taken from reference [15]. Units = moles. Where two figures are given they refer to Goldschmidt/Horn, respectively

| Component | | | From | | Now | | |
|---|---|---|---|---|---|---|---|
| | | | Primary rock | Volatile | Air | 1 l. sea-water | Sediments |
| $H_2O$ | | | | 54·90 | | 54·90 | |
| Si | $(SiO_2)$ ... | ... | 6·06/12·25 | | | | 6·06/12·25 |
| Al | $(AlO_{1\cdot5}, Al(OH)_3)$ | ... | 1·85/3·55 | | | | 1·85/3·55 |
| Cl | $(HCl)$ ... | ... | 0·01/0·02 | 0·54/0·94 | | 0·55 | —/0·40 |
| Na | $(NaO_{0\cdot5}, NaOH)$ | ... | 0·76/1·47 | | | 0·47 | 0·29/1·00 |
| Ca | $(CaO, Ca(OH)_2)$ | ... | 0·56/1·09 | | | 0·01 | 0·55/1·08 |
| Mg | $(MgO, Mg(OH)_2)$ | ... | 0·53/0·87 | | | 0·05 | 0·48/0·82 |
| K | $(KO_{0\cdot5}, KOH)$ | ... | 0·41/0·79 | | | 0·01 | 0·40/0·78 |
| C | ... ... ... | ... | 0·02/0·03 | 0·60/2·06 | | 0·002 | 0·62/2·09 |
| | $(CO_2)$ ... ... | ... | 0·02/0·03 | 0·53/1·05 | | 0·002 | 0·55/1·08 |
| | $(C(s))$ ... ... | ... | | 0·07/1·01 | | | 0·07/1·01 |
| $O_2$ | ... ... ... | ... | | 0·027/0·022 | 0·027/0·022 | | |
| Fe | ... ... ... | ... | 0·55/0·91 | | | | 0·55/0·91 |
| | $(FeO, Fe(OH)_2)$ | ... | 0·32/0·53 | | | | 0·18/0·32 |
| | $(FeO_{1\cdot5}, FeOOH)$ | ... | 0·23/0·38 | | | | 0·37/0·59 |
| Ti | $(TiO_2)$ | ... | 0·06/0·12 | | | | 0·06/0·12 |
| S | ... ... ... | ... | 0·01/0·02 | 0·06/0·06 | | 0·03 | 0·04/0·05 |
| F | $(HF)$ ... ... | ... | 0·03/0·05 | | | | 0·03/0·05 |
| P | $(PO_{2\cdot5}, H_3PO_4)$ | ... | 0·02/0·04 | | | | 0·02/0·04 |
| Mn | $(MnO_{1 \text{ to } 2})$ ... | ... | 0·01/0·05 | | | | 0·01/0·05 |
| $N_2$ | ... ... ... | ... | | 0·101/0·082 | 0·101/0·082 | | |

Table 6.2. Sillén suggests that these equilibriate and buffer everything in Table 6.3 at more or less constant values, i.e. not only a pH stat but a $pNa^+$ stat, $pMg^{2+}$ stat, etc.

TABLE 6.3.  The main constituents (and some others) in moles/kg. Sea-water ($\sim$0·975 l.) at 3·5% salinity (=35 g. dissolved salts/kg. sea-water). Taken from reference [15]

| | | | | | |
|---|---|---|---|---|---|
| $H_2O$ | 53·557 | $Sr^{2+}$ | 0·0001 | $F^-$ | 0·0001 |
| $Na^+$ | 0·4680 | $Cl^-$ | 0·5459 | $B(OH)_3$ | 0·0004 |
| $Mg^{2+}$ | 0·0532 | $SO_4^{2-}$ | 0·0282 | $N_2$ | 0·0006 |
| $Ca^{2+}$ | 0·0103 | $HCO_3^-$ | 0·0023 | $O_2$ | 0·0004 |
| $K^+$ | 0·0099 | $Br^-$ | 0·0008 | $NO_3^-$ | $0-35 \times 10^{-6}$ |
| | | | | $HPO_4^{2-}$ | $0-2·3 \times 10^{-6}$ |

TABLE 6.4. Present amount of ions in sea and amount of dissolved ions added by rivers in 100 million years in moles/cm² of total earth surface. Taken from reference [15]

| | $Na^+$ | $Mg^{2+}$ | $Ca^{2+}$ | $K^+$ | $Cl^-$ | $SO_4^{2-}$ | $CO_3^{2-}$ | $NO_3^-$ |
|---|---|---|---|---|---|---|---|---|
| Present in ocean | 129 | 15 | 2·8 | 2·7 | 150 | 8 | 0·3 | 0·01 |
| Added in 100 m. years | 196 | 122 | 268 | 42 | 157 | 84 | 342 | 11 |

The relationship governing this postulated titration is the 'geochemical balance equation'

$$\text{igneous rocks} + \text{volatiles} \leftrightarrow \text{sea-water} + \text{sediments} + \text{air}$$

The constituents of each of these five groups are listed in Table 6.2. It is important to note that this is really the equation of an acid–base titration, the acids being the volatiles and the bases the igneous rocks. Before we can do any calculations we need to know the quantity of titrant being added, i.e. what flows out of our rivers. These titrant details are given in Table 6.4. Other relevant quantities are listed under *from* in Table 6.2 or are collected from solubility products, formation constants and partial pressure constants given in the literature [1]. HALTAFALL may now be instructed to calculate the equilibrium ionic composition of 1 l. of these ingredients and the output is pH=8·1 and other concentrations as under *now* in Table 6.2. Thus the continuous titration hypothesis explains the present composition of sea-water but how about the 'stat' property? To prove that the ionic composition is buffered constant we have to calculate what happens to the equilibrium when $H^+$ or $Na^+$ or $Mg^{2+}$ are added. In all cases there is a readjustment of the amounts of solids present but the resulting ionic composition remains constant. Hence the hypothesis explains the real system.

In order to see *why* such a hypothesis works we need to simplify our system and study a model consisting of a few of the major components of the real system.

Consider the HCl, $H_2O$ (acids), $SiO_2$, $Al(OH)_3$, KOH (bases), five component system. There are three possible solid phases, $SiO_2$ (quartz), $Al_2Si_2O_5(OH)_4$ (kaolinite) and $KAl_3Si_3O_{10}(OH)_2$ (potassium mica), and three ions in solution $H^+$, $K^+$ and $Cl^-$, and the only gas above the solution is $H_2O$ vapour (Fig. 6.11). The

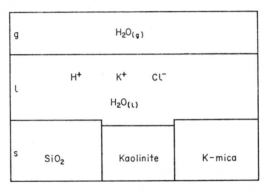

FIG. 6.11. Equilibrium in a five component system with gas, solution and solid phases present. From Sillén [15].

computers indicate that this is a pH stat as long as the $[Cl^-]$ and temperature $T$, remain constant. The phase rule predicts this or even more simply we can see this from:

$$1\cdot5 \text{ Kaolinite(s)} + K^+ \rightleftharpoons K \text{ mica(s)} + 1\cdot5 H_2O + H^+$$

Solids have activities $= 1$ so $\beta = [H^+]/[K^+]$. Furthermore, a litre of sea-water does not have a residual electric charge so:

$$[H^+] + [K^+] = [Cl^-] + [OH^-]. \text{ Actually, } [OH^-] = K_w/[H^+]$$

$\beta$ and $K_w$ are temperature dependent and if we hold $[Cl^-]$ constant we effectively have two equations in the two unknowns, $[K^+]$ and $[H^+]$. These two equations can only be solved to give unique values and so there will be a definite pH and $pK^+$ for each $[Cl^-]$ chosen.

Furthermore, it does not matter how much acid or alkali we add to the system, the pH remains constant until one of the solid phases is completely removed and as long as the [Cl$^-$] is constant.

Obviously this simple model suggests that we ought not to run fresh water into the oceans or to add sodium chloride as this will change the pH. However, if we add four more components to the model (NaOH, MgO, CaO and CO$_2$) it can be demonstrated that we now have pCl$^-$ stat as long as $T$ and [Mg$^{2+}$] are held constant. To obtain a pMg$^{2+}$ stat we must add more components, etc. In fact, the number of constituents required to hold everything more or less constant is 17, the 17 listed in Table 6.2. We write more or less because (a) there are many more minor constituents in the sea—in 1963 Sillén listed 54 elements as being definitely present and all these must play a small part, and (b) we have considered an equilibrium model but the sea is never really at equilibrium because of stirring, temperature gradients (25°C at surface, 5°C at 600 m. deep) and deficiencies (silica eating algae lower the surface silica concentrations). However, one cannot help remarking upon the super-efficiency of nature in arranging all these equilibria and upon the usefulness of HALTAFALL and suchlike in helping one to examine nature's equilibria.

## REFERENCES

[1] SILLÉN, L. G. and MARTELL, A. E., *Stability Constants of Metal Ion Complexes*, Spec. Publication No. 17, 2nd edn (The Chemical Society, London, 1964) (3rd edn in preparation).

[2] PERRIN, D. D., *Dissociation Constants of Organic Bases in Aqueous Solution* (Butterworths, London, 1965).

[3] ÖSTERBERG, R., *Ark. Kemi*, **25**, 177 (1966); and *Fil Doc. Thesis* (Gothenberg, 1966).

[4] DYRSSEN, D., JAGNER, D. and WENGELIN, F., *Computer Calculation of Ionic Equilibria and Titration Procedures* (Wiley, 1968), New York and London.

[5] BIEDERMANN, G., *Svensk Kemisk Tidskrift*, **76:4**, 1 (1964).

[6] ROSSOTTI, F. J. C. and ROSSOTTI, H., *The Determination of Stability Constants* (McGraw-Hill, London, 1961).

G

[7] INGRI, N. and SILLÉN, L. G., *Ark. Kemi*, **23**, 97 (1964).

[8] SAYCE, I. G., *Talanta*, **15**, 1397 (1968).

[9] WILLIAMS, D. R., *J. chem. Soc.*, A 1550 (1970).

[10] ANDREWS, A. C. and GRUNDEMEIER, E. W., *J. inorg. nucl. Chem.*, **28**, 455 (1966).

[11] PERRIN, D. D. and SAYCE, I. G., *Talanta*, **14**, 833 (1967).

[12] INGRI, N., KAKOŁOWICZ, W., SILLÉN, L. G. and WARNQUIST, B., *Talanta*, **14**, 1261 (1967).

[13] PERRIN, D. D., *Suomen Kemistilehti*, **42(9)**, 205 (1969).

[14] SILLÉN, L. G., *Svensk Kemisk Tidskrift*, **75**, 4 (1963).

[15] SILLÉN, L. G., *Chemy Brit.* 291 (1967).

*Chapter 7*

# CALORIMETRY

---

WHEN THE scientific historians write the history of the 20th century, the lack of progress in applying calorimetry to living molecules will no doubt be mentioned. Consider the four simplest properties of any compound that we need to know—its size, its shape, the three-dimensional layout of its atoms, and the strengths of the bonds involved. Methods are well established for determining the first three of these—we assume size approximates to weight and use the ultracentrifuge, we merely look at the molecule to decide its shape (albeit under a different light in the electron microscope), and we bounce X-rays off the atoms to determine their three-dimensional configurations. These three approaches have all been discovered, developed and rapidly applied to living systems. Furthermore, when a bond is made or broken, energy is involved and this exhibits itself in the form of heat. Hence we have only to calorimetrically measure this heat to measure the bond strengths and to follow the extent of any chemical reaction. This method was known and applied more than a century ago and yet is still not generally applied to living systems. Fortunately, moves are now being made in this direction, and, indeed, not before time. Many laboratory techniques are much newer and already more advanced than calorimetry and these often involve upsetting the molecule by distortion, excitation or even annihilation (e.g. nuclear magnetic resonance (n.m.r), electron paramagnetic resonance (e.p.r.), infrared (i.r.), optical rotatory dispersion (o.r.d.), and mass spectra (m.s.)).

Restricting our discussion to solution chemistry, calorimetry may be used (a) in instantaneous reactions to measure $\Delta H$, the bond strengths, directly, and also $\Delta S$ indirectly if $\Delta G$ be known; (b) to follow the rate of a slow biological reaction; (c) as a method of assay.

### 7.1 *Calorimetry of reactions which are complete within the time of mixing*

For an isothermal process [1]

$$\Delta G^0 = -RT\ln\beta = \Delta H^0 - T\Delta S^0$$

Considering equation (6.1)

$$p\text{AH}(H_2O)_w + q\text{B}(H_2O)_x \rightleftharpoons A_pB_qH_r(H_2O)_y + (p-r)\text{H}(H_2O)_z + (w+x-y-z)H_2O$$

and assuming that the $\beta$ for this reaction has been measured, calorimetrically measuring $\Delta H$ will give us the $T\Delta S$ term.

$\Delta H$, the enthalpy change, combines several factors, the heat of complex formation, the heat of removing $(w+x-y-z)H_2O$ from the aquated protonated ligand and from the metal ion, the heat of deprotonation of the ligand and the heat of aquation of any protons released from the ligand, $(p-r)\text{H}$.

$\Delta S$, the entropy change, also has several contributory factors— the change in the number of particles ($\sum$products–$\sum$reactants), steric strains in $A_pB_qH_r(H_2O)_y$ as product and any strains relieved by destroying the reactants.

Additionally (see equation 6.2), heats and entropies arise from hydrolysis of reactants and products.

Ideally it is preferable to have heats and entropies for each step of

$$A + B \rightleftharpoons AB \overset{A}{\rightleftharpoons} A_2B \overset{B}{\rightleftharpoons} A_2B_2 \ldots A_pB_q, \text{ etc.}$$

but the values of $w$, $x$, $y$ and $z$ are not known so we must be content to have to deal with hydrated species throughout. This means that heats and entropies of dehydration 'pollute' our results and must not be forgotten when discussing enthalpies and entropies of solution reactions.

7.1.1 *Choice of medium*   Heats observed for $I=0$ may be related directly to bond strengths. Unfortunately as $I \to 0$, $Q_{meas} \to 0$ joule.

The greater the quantity of product we can produce, the larger $Q_{\text{meas}}$. Hence the medium considerations of Chapter 6 apply to calorimetry bearing in mind that (a) the hydration spheres in 3 M $ClO_4^-$ will not be the same as at $I=0$, (b) $\Delta H^0$ is not absolutely equal in value to $\Delta H^{3\,M}$, and (c) that entropy changes are not so evident in high $I$ media [2].

Clearly we must aim for methods usable at low $I$ values and this means improved calorimetric techniques and equipment. However, even though $\Delta H$ values in ionic media are not absolute, we do expect to observe the same trends at $I=3\,M$ as at $I=0$ and it is these trends that are useful for inorganic-biological research. Other standard conditions used are 1 atm pressure and $25\cdot0$ or $37\cdot0°C$.

Two concluding remarks: Christensen and Izatt list equations for adjusting $\Delta H^I$ to $\Delta H^0$ values. These latter are used by physical and theoretical chemists. The van't Hoff equation method of plotting log $K$ versus $1/T$ also gives $\Delta H$ and $\Delta S$ values but in complicated biological systems the method is of doubtful accuracy [3].

7.1.2 *Titration calorimetry* [3]  The minimum number of experimental measurements necessary to solve for $p$ unknown heats of reaction is $p$. In practice many more points are usually obtained experimentally by employing a titration procedure and these are then solved by a least squares approach. This titration can be of two types, *continuous* or *incremental,* and these methods are represented diagrammatically in Fig. 7.1. In the continuous method the titrant is added continuously to the titrate until the reaction is completed. In the incremental method small amounts of titrant are added at about 30 minute intervals and each point has its own fore and aft period. Often between the aft period of one point and the fore period of another, cooling or heating adjustments to the temperature of the titrate may be made. This ensures that all measurements are made at exactly the same temperature. The continuous method is completed within a couple of hours and so several titrations may be performed during a day whereas the incremental method takes a day per titration and is slightly more accurate as (a) all the results are at one temperature, and (b) there is no time lag associated with each point plotted on the thermogram. (In the continuous method the heat observed is always a

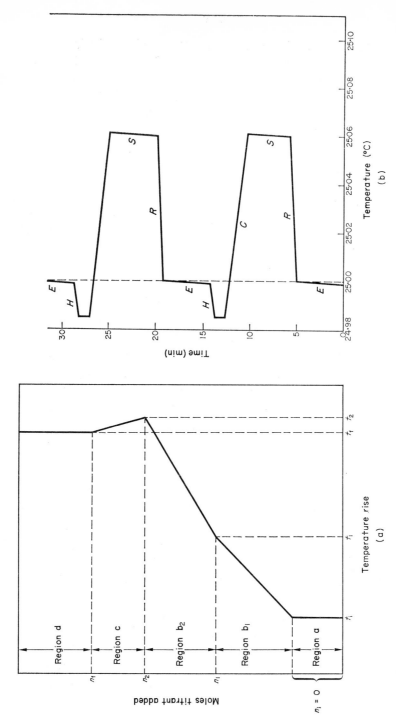

Fig. 7.1. Examples of (a) continuous and (b) incremental thermograms, from Christensen and Izatt [3]. (a) Titration of HCl/CH₂ClCOOH solution with NaOH: regions a and d—no titrant added; $n_1$ =end point for HCl/NaOH reaction; $n_2$ =end point for CH₂ClCOOH/NaOH reaction; region c represents heats of adding titrant and this correction ought to be applied to regions $b_1$ and $b_2$. (b) Typical incremental titration calorimetry thermogram with temperature adjustment. $E$ =equilibrium to thermostat temperature, $R$ =reaction heat, $S$ =equilibrium period after reaction, $C$ =cooling period, and $H$ =heating period.

minute or so behind the true mass balance and $\beta$ conditions inside the calorimeter vessel—due to the kinetics of mixing and the heat flow into its surroundings and the detector probe.) Whichever the method, one obtains a thermogram and it is instructive to compare this with a percentage species distribution plot (Fig. 7.2).

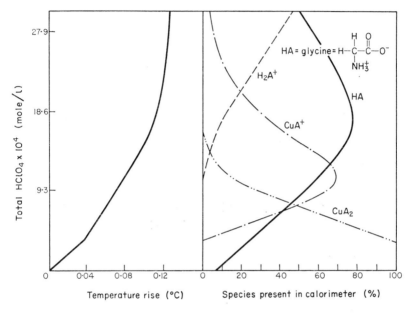

FIG. 7.2. Correlation of temperature rise with percentage species present in calorimeter for the titration of copper (glycinate ion)₂ with HClO₄. (Reproduced by permission of *Inorg. Chem.*, **3**, 1565 (1964). Copyright 1970 by the American Chemical Society. (Reprinted by permission of the copyright owner.)

This plot may be calculated from HALTAFALL or COMICS using known formation constants. The two curves give the following useful information:

1. The slopes of the thermograms give $\Delta H_1$, $\Delta H_2$, etc. (see later).
2. An inflection point in the thermogram gives a check on the stoichiometry of the system if this point corresponds to an inflection in the species distribution plots.

The actual shape of the thermogram is governed by several factors.

(a) The number of reactions involved gives the maximum number of possible end points discernible.

(b) The relative equilibrium constants decide whether breaks are sharp rounded or just slight inflections. In fact a sharp break pre-requires about six log units between consecutive formation constants.

(c) The relative enthalpies of formation for the complexes formed determine the slope of the thermogram and whether it increases or decreases after an inflection point, e.g. if $-\Delta H_2 > -\Delta H_1$ the slope decreases.

(d) Heat corrections for the heats of dilution of titrant and titrate, heat losses to the surroundings and temperature differences between titrant and titrate all affect the correction slope.

For example, in the copper glycinate titration shown in Fig. 7.2, the overall pattern of the thermogram results from the following equilibria.

$$NH_3^+.CH_2.COO^- \rightleftharpoons NH_2.CH_2.COO^- + H^+$$
$$NH_3^+.CH_2.COO^- + H^+ \rightleftharpoons NH_3^+.CH_2.COOH$$
$$H^+ + OH^- \rightleftharpoons H_2O$$
$$NH_2.CH_2.COO^- + Cu^{2+} \rightleftharpoons Cu(NH_2.CH_2.COO)^+$$
$$Cu(NH_2.CH_2.COO)^+ + NH_2.CH_2.COO^- \rightleftharpoons Cu(NH_2.CH_2.COO)_2$$

### 7.1.3 Converting thermograms into heats of reaction values

Most calorimeters use electrical calibration employing a small heating coil immersed in the titrate. A known amount of electrical energy in joules, accurately measured by passing a known current for a known time through a known resistance, gives a temperature rise and this may be compared directly with the temperature change from the chemically generated heating. Chemical calibration may also be used, usually from the well-established heat of formation of water from its ions, heat of dissolution of potassium chloride crystals or heat of protonation of tris(hydroxymethyl)aminome-thane (THAM). The highest accuracy is attained when the calibra-tion uses the same volume in the vessel and gives the same number of joules released over the same time interval as the heat evolved during the complexing reaction.

An incremental titration, by definition, is a series of steps or points. A continuous titration may also be considered as a series of equally spaced points (by, for example, striking off 1 ml intervals on the thermogram). For each point, by either method, we have the number of joules recorded, $Q$, and the change in the number of moles of each complex species present, $\Delta n_n$. If $\Delta H_n$ is the heat of complex formation for the nth complex, the first and second points give rise to equations of the form

$$Q_1 = (\Delta H_1 . \Delta n_1)_1 + (\Delta H_2 . \Delta n_2)_1 \ldots (\Delta H_p . \Delta n_p)_1$$
$$Q_2 = (\Delta H_1 . \Delta n_1)_2 + (\Delta H_2 . \Delta n_2)_2 \ldots (\Delta H_p . \Delta n_p)_2$$

and so on to

$$Q_r = (\Delta H_1 . \Delta n_1)_r + (\Delta H_2 . \Delta n_2)_r \ldots (\Delta H_p . \Delta n_p)_r$$

When $r = p$, the equations can be solved uniquely. When $r \gg p$, as is usual in practice, a least squares set of 'best' $\Delta H$ values is electronically computed using LETAGROP to find $U_{min}$. for the error square sum [4]

$$U = \sum_{i=1}^{r} (Q^i{}_{measured} - Q^i{}_{calculated})^2$$

### 7.1.4 Determination of formation constants during the calorimetry

There are two methods of determining the formation constants.

1. If the system being investigated lends itself to pH electrode work the simplest and most accurate method merely involves having a glass/reference electrode combined probe inside the calorimeter (see experimental section) and using the incremental method [5]. The approach is then equivalent to that outlined in Chapter 6 and has the advantage that it immediately checks our HALTAFALL calculations of the conditions inside the reaction vessel. There are two small disadvantages with this method: (a) the electrodes tend to increase the amount of heat leaking from the vessel, and (b) the calomel reference electrode is not always acceptable as it leaks chloride ions.

2. Entropy titration approach [3]. $\beta(\Delta G)$, $\Delta H$ and $\Delta S$ can sometimes be determined from calorimetric data alone provided that the $\beta$ values are not too large, i.e. all the reactions do not go quantitatively to completion.

Consider  $B + A \rightleftharpoons BA$
$\qquad B + 2A \rightleftharpoons BA_2$
$\qquad B + rA \rightleftharpoons BA_r$

For any point, if $[BA_r] = 0$ at the beginning of the titration

$$Q = \sum_1^r ([BA_r] \Delta H_r . V_r) \qquad (7.1)$$

where $V_r$ is the volume in litres

$$\beta_r = \frac{[BA_r]}{[B][A]^r} \qquad (7.2)$$

$$T_A = [A] + [BA] + 2[BA_2] \ldots + r[BA_r] \qquad (7.3)$$

$$T_B = [B] + [BA] + [BA_2] \ldots + [BA_r] \qquad (7.4)$$

These four relationships may be combined to give one equation in $2r$ unknowns ($r$ unknown $\beta_r$ values and $r$ unknown $\Delta H_r$) and as long as $2r$ or more points are measured ($s$, say) the equation is theoretically solvable. Computational methods have been devised in which values of $\beta_r$ are guessed, equations (7.2), (7.3), and (7.4) are solved for each guess to give $[BA_r]$ and then $\Delta H_r$ are obtained from equation (7.1). A pattern of guesses is tried until the 'best' set of $\beta_r$ and $\Delta H_r$ values is found. The least squares criteria for choosing the best set are defined as $U_{\min}$ when

$$U = \sum_{n=1}^s (Q_n - V \sum_{m=1}^r [BA_m] . \Delta H_m)^2$$

Clearly the arguments do not apply if some $\Delta H = 0$.

7.1.5 *Experimental* Tyrrell and Beezer [6] describe a host of different varieties of solution calorimeters and these details will not be repeated here. We shall describe a solution calorimeter in common use and then a new variety of microcalorimeter.

*Solution calorimeter.* The calorimeter shown in Fig. 7.3 is basically that of Gerding, Leden and Sunner. It consists of an inner reaction vessel and an outer shielding vessel that together are totally immersed in a thermostat bath. The inner surface of the reaction vessel and the lower side of its lid are gold plated. All other surfaces are coated in shiny nickel. Four probes (and sometimes an optional fifth) are sealed into the top of the inner vessel.

The *heater* is used for electrical calibration and is the known

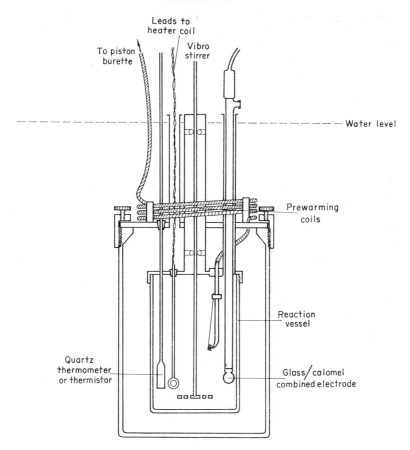

Fig. 7.3. A typical isothermal solution calorimeter.

resistance through which a known current is passed for a known time. These three variables are measured by the electronics of the control monitor. The heater consists of 1 metre of manganin wire (*c.* 25Ω), non-inductively wound and immersed in oil in a gold thimble. More recently, shorter lengths (5 cm) of higher resistance wire are used bare and are just coated with a layer of chemically resistant epoxyresin. This reduces heat losses up the stem of the support holding the gold thimble. Although more time-consuming, chemical rather than electrical methods of calibration are preferable as they do not involve heat losses up the wires of the leads to any heater.

The *burette tip* is of glass or gold and has a self-closing device that prevents back diffusion. Careful design is necessary to ensure that the temperature of the titrant passing through the tip is as near as possible to that of the titrate in the vessel. This is achieved by either having a prewarming spiral between inner and outer vessel, in which case a heat sink is needed to save robbing the reaction solution of heat (a pot of mercury or Wood's metal around the spiral may be used for this) or the spiral may be placed in the thermostat bath, on top of the outer vessel, and then the bath acts as the heat sink. A matched pair of temperature sensors (e.g. thermistors), one in the vessel and one in the end of the spiral, may be used to check and measure the temperature difference between titrant and titrate. The free end of the spiral is joined to a thermostatted piston burette.

The *thermistors* (*therm*ally sensitive re*sistors*) are small semi-conductors that are sealed in the ends of glass tubes and have the fortunate property that their resistance is temperature dependent. Their resistance decreases as temperature increases and, over the narrow range we are interested in, the resistance may be said to be proportional to the temperature. The thermistor is incorporated into one arm of a Wheatstone bridge and resistance readings versus time are noted and plotted to give Fig. 7.1. Alternatively, if we are comparing the temperature of the titrant with titrate the resistance box (Fig. 7.4) is replaced by the second thermistor. With matched thermistors, when the galvanometer registers zero deflection, the titration may be commenced.

Vibro *stirring* is preferable to paddle stirring as it reduces the heat of stirring correction, i.e. the fore ($E$) and aft ($S$) slopes in Fig. 7.1 (b) become more horizontal. Also, this is necessary in reactions where there is a tendency for precipitation (all too frequently in biological systems!) unless the titrant is rapidly removed away from the burette tip. If this localized precipitation does occur it (a) will have an enthalpy effect and (b) may be slow to redissolve. Hence efficient vibro stirring is an asset.

Optional *extra probe*. It is often advantageous to know the pH inside the solution so a combined glass/calomel reference electrode of the slim variety may be incorporated as an extra probe. The results from such an electrode pair may be used to compute the concentrations of the species inside the calorimeter. However, this probe means that a large volume of glass and electrode solution is being introduced and this may lead to a small time lag in temperature response.

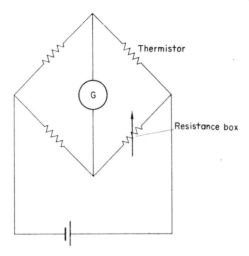

FIG. 7.4. A simple Wheatstone bridge circuit for measuring thermistor resistance. G is a galvanometer or null point detector.

The calorimeter just described is of the constant environment type so a thermostatted bath ($\pm 0.001°C$) and room ($\pm 0.1°C$) is necessary. The instrument is designed to study reactions that, for each point, reach equilibrium within the time of mixing and, at any rate, within 30 min of commencing the addition. For these slower reactions, evacuating the air space between the inner and outer vessels increases the precision. The specifications of the calorimeter are: titrate = 15–100 ml, titrant = 0–25 ml, aliquot = 0–2.5 ml, heat measured 1–30 $\pm 0.02$ cal (4–120 $\pm 0.09$ joules) and sensitivity = $\pm 5 \times 10^{-5}°C$.

## 7.2 *Microcalorimetry* [7]

The titration calorimeter that we have just discussed depends upon the fact that the heat passing in or out of the calorimeter from the heat sink (i.e. from the thermostat bath and outer vessel) moves so slowly that the fore and aft periods are linear, i.e. any heat generated is assumed to stay in the solution in the inner vessel and its immediate environment. It does, however, slowly escape and this is the limiting factor to the length of time that each experimental point may occupy. Microcalorimetry uses the converse principle—it encourages the heat to escape but insists that it does so only via a thermocouple arrangement that summates the heat passing through it. Thus the reaction vessel must be surrounded by a large heat sink, there being very many thermocouple connections between the vessel and the sink and the minimum number of other connections entering the vessel as the heat must not escape uncounted. The total amount of heat involved depends upon the quantities of chemicals present and this calorimeter is best suited for micromole quantities or millimolar solutions. These are ideal for studying the complex world of living matter.

A commercial microcalorimeter [8] is shown in Figs 7.5 and 7.6.

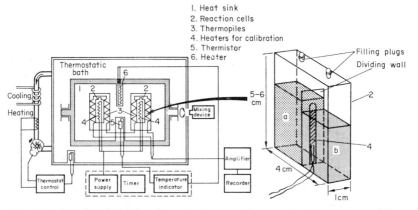

FIG. 7.5. The LKB Batch microcalorimeter. The reaction is commenced by rotating the whole heat sink about the horizontal axis so that reactants a and b are mixed together.

Heat, unlike chemicals, cannot be contained in a test-tube and in order to channel heat loss or gain through thermocouples, thousands are needed. The reaction vessel is actually surrounded by a batch of thermocouples known as a thermopile. A single thermocouple is merely two *different* metal wires joined together in a loop. If one metal–metal junction is warmer than the other a potential difference exists between them and a current flows. If we

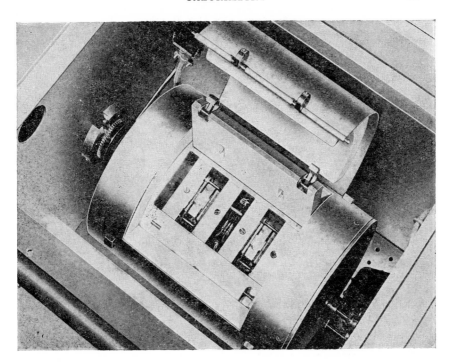

FIG. 7.6. Birds-eye view of a batch microcalorimeter. The lid of the heat sink is hinged open to expose the two reaction cells, each with two filling caps uppermost. The reaction is commenced by rotating the whole heat sink drum about an axis running from the top left-hand corner of the photograph to the bottom right-hand corner. (By courtesy of LKB Instruments Ltd.)

have $n$ thermocouples all connected in series and all having one end touching the reaction vessel and the other touching the heat sink, heat will escape along these wires and $n \times$ potential difference will be the voltage observed. For a given total amount of energy released, the instrument can either be designed to give a long lasting weak signal or a short strong signal (called the heatburst approach). This latter approach is aimed for because it minimizes disturbances from environmental changes.

Heat corrections are necessary for friction of vessel movements, heats of mixing (mixing may be performed by rotating the whole calorimeter unit about its central axis), and for changes in the temperature of the heat sink. The simplest method of applying these corrections is to use the twin principle in which *two* vessels are used, one charged with reactants and the other with a non-reacting liquid. The thermopiles from these two are connected in tandem in opposition so that the correction from the blank vessel automatically subtracts itself from the potential of the reaction vessel. Naturally, chemical heat corrections also have to be applied just as for the calorimeter previously described. The heat-burst apparatus is ideal for studying instantaneous reactions of small sample volumes (batches of up to 2 ml added to up to 4 ml). For investigating fast reactions ($<1$ min) on larger volumes the apparatus may be converted into a flow microcalorimeter in which case the reaction vessels are replaced by vessels where two streams of reactants mix. For slower reactions, the reactants may be mixed outside the vessels and the product solution (or more pedantically, the mixture of reactants that are slowly reacting to form products) is pumped through the cell. Under these circumstances most of the reaction takes place outside the vessel and the calorimeter is being used as a sampling device to measure the heat evolved per second over the several hour course of the reaction.

The calorimeter may be calibrated chemically or by a small internal heater and the batch method is capable of measuring $4 \times 10^{-3}$ joules with an accuracy of $\pm 4 \times 10^{-7}$ J. The flow calorimeter has a maximum flow rate of 20 ml $h^{-1}$ and a sensitivity of $4 \times 10^{-7}$ J $s^{-1}$.

### 7.3 *Uses of calorimetric results*

Many examples of the uses of calorimetry appear in the literature and just a few representative ones will be given here. (The HSAB uses of $\Delta H$ and $\Delta S$ have already been mentioned.)

*Example 1.* Conventional solution calorimetry can be used for unravelling the stereochemistry of complexes in solution and this may be illustrated by studying the divalent transition metal ion complexes of histidine and of tryptophan for comparison purposes [9]. Up to pH 5, histidyl ligands have been shown to be bidentate and protonated on their primary amine groups. Thereafter the structures of the complexes are unknown. Crystallographic determinations of the *solid* complexes have indicated that the zinc histid*ine* complex is tetrahedral [10], [11]. In solution there is now evidence that the zinc histid*yl* complexes have bidentate ligands (one bond to the imidazole and another to the primary amine nitrogen), are tetrahedral, and that the unprotonated carboxyl group hangs free. Reasons for suggesting this postulate come from the formation curves in Fig. 7.7 and the thermo-

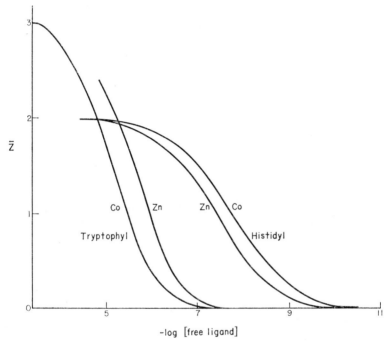

FIG. 7.7. Formation curves for cobaltous and zinc histidyl and tryptophyl complexes. $\bar{Z}$ is the average number of ligands attached to each metal ion. Taken from Williams [9].

dynamic parameters listed in Table 7.1. Histidine, in fact, is the only amino acid that has its zinc complexes less stable than its cobalt [12]. Evidence for a tetrahedral unprotonated complex of zinc in solution is 1. recalculation of the histidyl–zinc formation curves on the basis that a *protonated* carboxylic group hangs free gives non-superimposable curves. Hence the group must be unprotonated, i.e. —$COO^-$; 2. $\Delta S$, for the formation of the zinc–histidine is higher than expected: if the six membered N—N ring

H

TABLE 7.1. Enthalpies and entropies for the formation of the 1:1 complexes of histidyl and tryptophyl with zinc (II) and cobalt (II). $I = 3.00$ M $ClO_4^-$, 25°C

| | Histidyl | | Tryptophyl | |
|---|---|---|---|---|
| | $-\Delta H_1°$ | $\Delta S_1°$ | $-\Delta H_1°$ | $\Delta S_1°$ |
| | kcal mol$^{-1}$; kJ mol$^{-1}$ | kcal K$^{-1}$ mol$^{-1}$; kJ K$^{-1}$ mol$^{-1}$ | kcal mol$^{-1}$; kJ mol$^{-1}$ | kcal K$^{-1}$ mol$^{-1}$; kJ K$^{-1}$ mol$^{-1}$ |
| Zn | 5·52   23·1 | 13·8   57·7 | 4·10   17·1 | 9·2   38·5 |
| Co | 5·63   23·5 | 15·2   63·6 | 2·78   11·6 | 11·6   48·5 |

needs a larger bond angle at the metal ion than the 90° found in octahedral complexes, the increased entropy could arise because when the complex becomes tetrahedral the strain in the ring is released; 3. $-\Delta H_1$ (zinc) is less than $-\Delta H_1$ (cobalt) because we are comparing the strength of two bonds from the ligand to zinc and three from the ligand to cobalt; and 4. The main reason for histidine being an anomalous amino acid, as far as its zinc complexes are concerned, arises because it has the possibilities of being tridentate whereas the other amino acids can usually only be bidentate.

*Example 2.* Christensen *et al.* [13] have studied the enthalpies of protonation of adenosine (a terminal group in soluble RNA)

of which it was already known that it had a pK above 12. First they used the entropy titration approach and the pK was found to be 12·35 ($I=0$ M, 25°C). This figure could not have been determined using a glass electrode because they do not respond well in strongly alkaline regions. Secondly, there was the question as to which one of the many possible sites actually claims the $H^+$ added. This is an important precursor to discussing how copper (II) is held to the molecule because it is believed that the site that holds the proton is also the one that holds the copper ion. Thus the pKs of series of N substituted adenosines were measured using the entropy titration approach and they were able to show quite clearly that it is the 1 position that accepts the proton.

*Example 3.* Microcalorimetry has many biological applications,

one of which is to follow bacteriological growth by flow calorimetry. A widely used non-calorimetric method for studying
reaction rates is that of turbidimetry but Fig. 7.8 clearly shows that

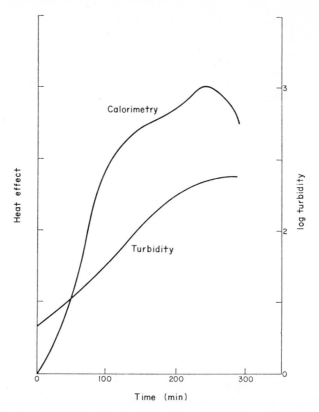

FIG. 7.8. A comparison of the flow microcalorimetric and turbidimetric
methods for following bacterial growth.

microcalorimetry gives a much more detailed curve. The system
being followed in the calorimeter is *Streptococci follicalis* growing
in a limited glucose medium [8]. The tremendous resolving power
of calorimetry may be further seen in Fig. 7.9 [14]. The growth
curve for bacteria *Escherichia coli* is reproducible in a second
experiment until an antibiotic is added after the first ten hours.
Then the reaction heat returns to the base line again. Clearly the
bacteriological applications of microcalorimetry are immense and

even more so if the problem of providing oxygen inside the reaction cell can be overcome (some bacteria cannot grow without an oxygen supply).

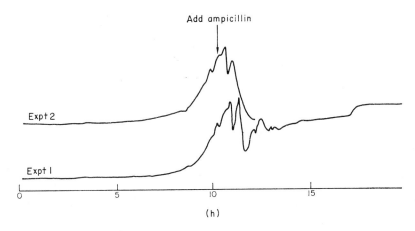

FIG. 7.9. The high resolving power of flow microcalorimetry is illustrated by following the growth of *Escherichia coli* and the effect of adding an antibiotic (ampicillin). From Wadsö [14].

*Example 4.* Assay by microcalorimetry. There are many examples in the literature [7]. The assay of horse serum cholinesterase may be taken as representative. Spectral methods give the activities of different amounts of pure enzyme and then a calibration curve of calorimetric response, $\Delta$, versus activity is constructed (Fig. 7.10). Unknown quantities of enzymes may be assayed by measuring their $\Delta$s and referring to the calibration curve. The conditions chosen were a flow rate of 20 ml h$^{-1}$, 13–15 min after the reactions had commenced. An additional technique is also incorporated in this assay method. It is called the 'amplification effect' and involves the cholinesterase assay measurements being performed in a tris(hydroxymethyl)aminomethane buffer. The enzymatic reaction releases protons which instantaneously react with the buffer to give a large heat effect. Conversely, if an excess of enzyme is used, the assay method may be employed as a means of assaying the substrate.

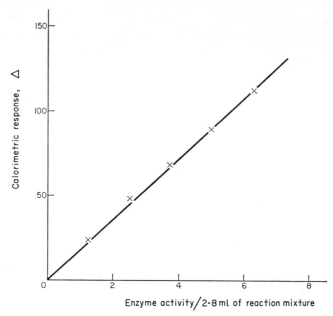

Fig. 7.10. Determination of cholinesterase activity. Calorimetric response, $\Delta$, is plotted against Rappaport units of enzyme activity. Amplification $= 10^4$. From Wadsö [7].

## REFERENCES

[1] WYATT, P. A. H., *Energy and Entropy in Chemistry* (Macmillan, London, 1967).

[2] AHRLAND, S., *Helv. chim. Acta*, **50**, 307 (1967).

[3] CHRISTENSEN, J. J. and IZATT, R. M., *Physical Methods in Advanced Inorganic Chemistry*, Ch. 11, p. 538 (Interscience, New York, 1968).

[4] GRENTHE, I. and WILLIAMS, D. R., *Acta chem. scand.*, **21**, 347 (1967).

[5] JONES, A. D. and WILLIAMS, D. R., *J. chem. Soc.* (A) (1970). In press.

[6] TYRRELL, H. J. V. and BEEZER, A. E., *Thermometric Titrimetry* (Chapman and Hall, London, 1968).

[7] WADSÖ, I. *et al.*, *Acta chem. scand.*, **22**, 927 and 1842 (1968), **23**, 29 (1969).

[8] *The LKB Microcalorimetry System*, LKB-Produkter AB, Fack, 161 25 (Bromma 1, Sweden).

[9] WILLIAMS, D. R., *J. chem. Soc* (A), 1550 (1970).

[10] FRASER, K. A. and HARDING, M. M., *J. chem. Soc.* (a) 415 (1967).

[11] HARDING, M. M. and COLE, S. J., *Acta crystallogr.*, 643 (1963).

[12] MARTELL, A. E. and SILLÉN, L. G., *Stability Constants of Metal Ion Complexes*, 2nd edn, Special Publication (The Chemical Society, London, 1964). 3rd edn in press.

[13] IZATT, R. M., RYTTING, J. H., HANSEN, L. D. and CHRISTENSEN, J. J., *J. Am. chem. Soc.*, **88**, 2641 (1966).

[14] WADSÖ, I., *Svensk kem. Tidskr.*, **6–7**, 28 (1969).

## FURTHER READING

BROWN, H. D. (Ed.), Biochemical Microcalorimetry (Academic, New York, 1969).

*Chapter 8*

# OTHER METHODS OF STUDYING METALS IN LIVING SYSTEMS

---

BIOLOGICAL RESEARCH is changing from the cellular to the molecular and so the chemist's tools for investigating the bonding in molecules can now be invoked. This chapter summarizes these methods and indicates their uses in unravelling the chemistry of life. Space will not permit us to discuss the underlying principles as deeply as we did for formation constants and enthalpies so the theories and mathematics is left for the experts who have written the reviews quoted in the references.

One ought not to be alarmed at the length of time required for applying some of these techniques. A decade or so ago, the structure determination of vitamin $B_{12}$ took ten workers ten years. However, as long as the requirements of applied biological research are permitted to dictate the direction of pure chemical research, methods of speeding up these analyses are soon found. The structures of subsequent vitamins occupy less people for less time [1], [2].

## 8.1 *Visible and ultra-violet absorption spectra* [3], [4]

These spectra originate from transitions occurring among the outer electrons of atoms, ions or molecules. In molecules, the electrons that are generally available for promotion and a corresponding absorption of energy are the $\pi$ electrons associated with double bonds. An electron transition requires (a) the correct wavelength of incident energy to promote the electron between one energy level and another (this corresponds to the wavelength of the absorption peak), and (b) the probability that the electron is free enough to be promoted. This depends upon how tightly the electron's orbital is coupled to the molecule as a whole. The so-called 'forbidden transitions' originate when this orbital is so firmly under the molecule's control that the incident energy is all dissipated as heat and no promotion occurs. Hence the wavelength of absorption indicates how difficult it is to promote an electron and the peak height is an indication of how many such electrons are promoted. Electronic and conformational changes in the complex exhibit themselves in the peak height and position. The subject is well treated in the literature and the relationships therein take into account many more ramifications than these just listed, e.g. the spectrum observed is really the *resultant* of the energy of promotion and the energy properties of electrons dropping back to their ground states by diverse routes.

## 8.2 *Optical rotatory dispersion and circular dichroism (o.r.d. and c.d.)* [5]

Optical activity has been known for many years and it is the variation of this optical activity with the wavelength of the incident light that gives o.r.d. and c.d. spectra. If one selects an optical enantiomorph and illuminates it with polarized light of a varying wavelength, $\lambda$, a plot of the angle of optical rotation, $\alpha$, versus $\lambda$ is an o.r.d. curve. Circular dichroism is defined as the difference in extinction coefficients for left-hand and for right-hand circularly polarized light ($\epsilon_l - \epsilon_r$). A c.d. spectrum is a plot of ($\epsilon_l - \epsilon_r$) versus $\lambda$. The o.r.d. and c.d. spectra are particularly interesting around the absorption peaks in the visible and ultra-violet (Fig. 8.1). They may be applied to correlations of configurational changes and can be used to follow the steric course of a reaction. Conformational changes and ion-pairing effects are also seen from o.r.d./c.d. analysis.

## 8.3 *Infra-red spectra* [6], [7]

Infra-red absorptions arise because of atomic and molecular vibrations and the deformation of bonds. Any forbidden absorption bands can usually be investigated by Raman spectra

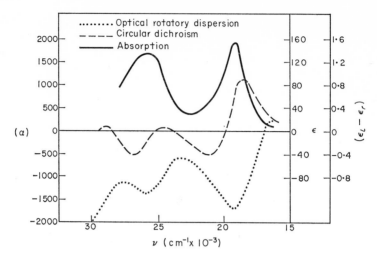

FIG. 8.1. Spectra of [Co(L-ala)₃]. Taken from Hill and Day [2].

(measurements of changes in wavelength of scattered mono-chromatic radiation).

Infra-red spectral results have two uses.

1. 'Fingerprinting' is possible because different kinds of bonds absorb in different wavelength areas. Libraries of infra-red finger-prints are now available but nevertheless a complete structure determination is usually impossible by this method alone and supplementary evidence from other methods is usually required. Bonds involving metal ions are found towards the far infra-red region, but the effects of metal–ligand bonds on *organic* bonds present in the ligand molecule may be observed in the near infra-red.

2. The height of an absorption peak is dependent upon the concentration of a bond present (and so, in theory, can be used to determine a formation constant), and the shift in a peak (e.g. such as occurs when a metal ion is added to a ligand) is related to the bond strengths present.

The work of Carlson and Brown [8] has already been referred to in Chapter 4 and their investigation of the distribution of the transition metal ion–histidine complexes will now be described as an illustration of the steps involved in investigating complexes in solution. The problem is to decide which donor atoms bind the metal ions and whether this binding changes with pH. Although potentiometry and calorimetry are more accurate ways of deciding *how many ligands* are attached to a metal ion under any given set of conditions and *how strong the bonds are*, such methods are not able to decide unambiguously *which donor atoms* are involved.

Unfortunately water, $H_2O$, has absorption bands in many of the infra-red regions that we wish to use so the following work uses deuterium oxide, $D_2O$, as solvent. Any ligand exhibits a spectrum which varies with pD because sometimes we have an acid form and sometimes a base form present. When the presence of a metal ion is introduced it is possible to measure quantitatively the effect which the metal ion exerts upon the pD dependence of the ligand spectrum. Further, this ligand spectrum is a collection of the characteristics of the protonated and of the deprotonated forms of different functional groups, e.g. $=N—$, $—NH_2$, $—COO^-$; hence whereas a potentiometric $\beta_{pqr}$ value says nothing directly about the state of the donor sites involved, under optimum conditions infra-red spectra give information about the proton equilibria at each donor site. Regretfully, the method uses rather high concentrations of ligand and metal ion ($c.\ 0\cdot2$ M) and does not yield formation constants with a high precision. On the other hand, if a series of metal ions of the same charge is studied, the order of their interactions with any given ligand may be determined moderately precisely.

A precursor to the histidine investigation was one of imidazole and of $\alpha$ alanine. By using the spectra of assorted deuterated derivatives of imidazole and of the imidazolium ion (a proton added to the pyridine nitrogen of imidazole) and a process of elimination and confirmation by reference to the literature, absorption wavelengths were assigned to different bonds. Many of these imidazole absorptions undergo sizeable shifts to higher frequencies (lower wavelengths) when the imidazole is protonated to form the imidazolium ion. This occurs because adding a proton or a metal ion to an imidazole ring causes an extensive $\pi$ electron redistribution. Assignments were also made for $\alpha$ alanine. Both imidazole and alanine exhibit shifts in their spectra as the pD is changed and it is possible, by measuring the extinction coefficients of selected peaks, to tabulate the pD value for the 'half reaction points' for the series of metal ions involved. These half reaction points correspond to the pD value at which the concentrations of protonated and deprotonated forms are equal. That this value is related to the formation constant for the metal complex is evident from

$$AH^+ \overset{K}{\rightleftharpoons} A + H^+$$
$$AH^+ + B^{n+} \overset{\beta_{11}}{\rightleftharpoons} AB^{n+} + H^+$$

and at pD (half reaction) $[AH^+] = [A]$. Further,

$$T_A = [A] + [AH^+] + [AB^{n+}],$$
$$T_B = [B^{n+}] + [AB^{n+}],$$
$$T_H = [H^+] + [AH^+],$$

and $K$ are known. Hence from five relationships and five unknowns, solve for $\beta_{11}$.

The order of the formation constants determined in this way obeyed the Irving–Williams series.

Band assignments for histidine were obtained by comparing its spectra with those of alanine, imidazole and 4-methyl-imidazole. Similarly pD (half reaction) values for histidine–metal ion complexes are reported for both imidazole coordination and amino nitrogen coordination. Hence, now that all the absorption bands are assigned it is easy to see which donor atoms are (a) unaffected (i.e. neither protonated nor complexed), (b) protonated, or (c) complexed, as pD changes. The results of this study are shown in Fig. 4.1.

### 8.4 *Electron paramagnetic resonance* (e.p.r.) [2] [4], [9]

Many species have unpaired electrons, e.g. atomic hydrogen, nitrogen, nitric oxide, free radicals and some transition metals and ions. These unpaired spinning electrons behave as small magnets and if subjected to a magnetic field they are inclined to align themselves with this field. However, if sufficient energy is pumped into the electron it may set itself against the field. The quanta of energy necessary could be seen as an absorption band. Alternatively, if we bombard the system under investigation with a reasonable amount of energy, e.g. a microwave beam, and then gradually increase the prevailing magnetic field, the field strength at which the electron just aligns itself with this field will be characteristic of the electron's environment (e.g. the other electrons present may be under Jahn–Teller conditions, there may be other unpaired electrons in the vicinity, and, of course, environment includes the ambient temperature). It is this environment dependence that makes e.p.r. an excellent structural probe.

### 8.5 *Nuclear magnetic resonance* (n.m.r.) [2]

Nuclei having odd numbers of either protons or neutrons (e.g. $_1^1H$, $_9^{19}F$, $_{15}^{31}P$) possess intrinsic magnetic moments just like the electrons of e.p.r. These nuclear magnets precess around the direction of the applied magnetic field and if then bombarded with radio frequency energy this precession becomes tilted against the field as the energy is absorbed. Thus identical nuclei irradiated at a fixed frequency would be expected to absorb energy at the same strength of applied field. In fact, the actual absorption peak depends upon the surrounding electrons and adjacent nuclei so, once again, the n.m.r. probe is an ideal tool for investigating electronic and atomic arrangements in molecules.

### 8.6 *Equilibrium studies using ion-selective electrodes* [10]

The uses fall into two spheres: we may monitor activities of *free* ions (a) *in vivo* to check that everything is 'normal', or (b) *in vitro* to obtain formation constants. For simple equilibrium studies of metal ligand interactions in which the ligand has a pK in the range 2–10, following the pH change with a hydrogen, glass or quinhydrone electrode is sufficient to establish both the pK and the formation constant for the metal complex. However, when ligands have inconvenient pKs or when polynuclear or hydrolysis complexes occur, it is desirable to supplement these measurements by using electrodes that are responsive to ions other than protons.

Several commercial ion-selective electrodes are now available; they have the design shown in Fig. 8.2, and use the membrane principle whereby a potential develops between two solutions of unequal concentration of the same ion separated by a membrane (e.g. in the glass electrode the membrane is a thin glass wall).

There are several variations on this theme: there are solid state electrodes in which the membrane consists of a single crystal or a disc of active material. Difficulties in forming a crystal or disc may sometimes be avoided by dispersing this active material in an

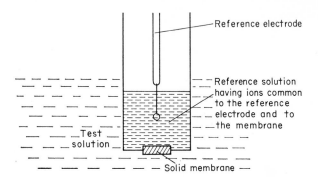

FIG. 8.2. General design of an ion selective electrode.

inert binder or matrix. Also there are liquid ion exchange membrane electrodes in which the ion of interest is attached to a large organic molecule of low water solubility.

The e.m.f. of such electrodes has been defined as:

$$E = \text{constant} + \frac{RT}{nF} \ln(a_j{}^n + K_{ij}a_j{}^n)$$

where $i$ and $j$ refer to the $n$ charged ion under consideration and an $n$ charged interfering ion. $K$ is called the selectivity constant. Even glass electrodes suffer this interference. The equation emphasizes that it is the relative activities of the ion under consideration and of the interfering ion that determines the selectivity, e.g. $Na^+$ ion interference to the hydrogen ion response of the glass electrode is negligible unless $a_{H^+} < 10^{-9}$ g ion 1.$^{-1}$. This type of interference can be tolerated if $K_{ij}$ values are known. More troublesome interference arises from poisoning, clogging or dissolution of the membrane. Nevertheless, if excessive interference means that the $E$ versus $\log a$ curve is not linear the electrode may still be used as a null point detector.

Commercial electrodes are available for the halide anions and for the non-transition cations found *in vivo*, $H^+$, $Na^+$, $K^+$, $Mg^{2+}$, and $Ca^{2+}$. The calcium ion electrode is very well known and has been applied to many bodily fluids. Other solid state and liquid ion exchange selective ion electrodes are also marketed but the transition metal ions are best monitored by amalgam electrode techniques. These have an excellent sensitivity and can be used to identify many complexes in solution [11] (see Fig. 8.3).

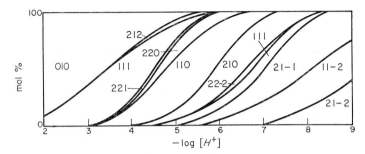

FIG. 8.3. This copper amalgam electrode analysis, at various pHs, of a solution of 10 mM copper (II) and 50 mM glycylglycylglycine (ggg) illustrates the sensitivity and usefulness of this technique. The regions are denoted by the numbers which are the subscripts to the general formula $A_pB_qH_r$. $A = ggg^-$, B = copper (II), e.g. 010 = free copper ions, 22–2 = $(ggg)_2Cu_2(OH)_2$. Taken from Österberg and Sjöberg [11].

## 8.7 Crystallographic studies by X-ray, electron and neutron diffraction [2], [12], [13]

The structure that crystallizes out from a solution usually bears some resemblance to those existing in solution. One ought not to reason in the converse direction and say that a postulated structure does not exist in solution because it cannot be obtained in the crystalline form. Crystallization and dissolution both involve different forces and bonds. By definition, the material under investigation must be crystalline and this crystal is treated as a three-dimensional diffraction grating since it consists of unit cells, of the order of 0·5–5·0 nm across, stacked in each direction. This crystal grating has been extensively used for diffracting monochromatic X-rays, because they have wavelengths similar in size to the interatomic distances, and more recently it is being used in electron and neutron diffraction studies as these give more accurate results. The X-rays are diffracted by the electrons present in the grating and the neutrons by the nuclei. Unfortunately, neutron diffraction requires (a) a source of neutrons and (b) larger crystals. Whichever method is used, the diffracted pattern gives a regular arrangement of spots on a photographic film. For X-rays, the cell dimensions determine the *directions* in which the beams are diffracted and the *intensities* of these beams depend upon the electron densities of the unit cells.

Modern diffractometers treat photographic plates as a thing of the past because visual estimations of spot intensities are usually only $\pm 10\%$. The four circle diffractometer uses electronic orientation of the crystals and the four angles defining the positions of the crystal and *electronic* counter, with respect to the incident beam, are, along with the reflection indices and intensities, punched on to tape or even fed directly into a computer. This plots the electron density map and can even draw the shape of the unit cell. In older methods the middle of the process involved the crystallographer 'selecting atom positions which are chemically sensible' and then the calculations were continued. The approach of the future is to remove this human element by allowing the computer to make an unbiased selection. The heavy atom method uses the principle that its reflections will dominate the diffraction pattern and so may be used as a marker or reference point in the calculations. The structure of vitamin $B_{12}$ was solved in this manner. For larger proteins, in order to resolve unambiguously the phase and structure relationships, several heavy metal atom derivatives are needed. Here again, metalloenzymes have been widely studied as it is possible to swop their metals for another metal. Problems encountered in protein crystallography are (a) they often contain a large unit cell of very many atoms, (b) trapped solvent and salts of crystallization may also pollute the diffraction pattern, and (c) the crystals may deteriorate after prolonged irradiation.

The usual approach is one of increasing resolution: one commences with low resolution (*c.* 0·6 nm) so that just the tertiary structure is observed, i.e. the coils of keratins or the globular shape of hemoglobin, and then increases the resolution to observe smaller entities. Structures have been determined for myoglobin, lysozyme, carboxypeptidiase, chymotrypsin, ribonuclease, papain and hemoglobin (*viz.* references within reference [12]). The last one listed was the epic work of Perutz and Kendrew which won the Nobel Prize in 1962—*The structure of hemoglobin.*

Hemoglobin (there is almost 1 Kg in each of us!) consists of four polypeptide chains and contains four iron (II) ions, each being capable of carrying an oxygen molecule from the lungs to the tissues. Degradation studies established the amino acid sequence of the 140 or so amino acid residues in each chain. X-ray crystallographic analysis gives the three-dimensional structure of the molecule although it does not reveal pertinent aspects about electronic factors since resolution below 1 nm is needed to make reasonable deductions about these. It arises that the oxygen is carried in a cleft in the hemoglobin. The gap between two chains opens to persuade the oxygen to enter, closes to carry and hold it and opens again to release the oxygen molecule. This concept was proved by a structure analysis on both the oxygenated and deoxygenated hemoglobin. The four chains consist of two $\alpha$ and two $\beta$ chains. They are folded as shown in Fig. 8.4 and fit snugly together as in Fig. 8.5, not only using the minimum

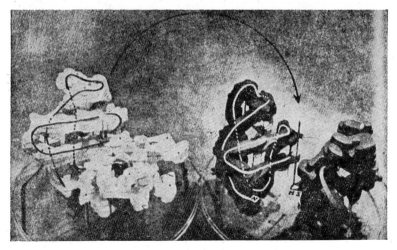

FIG. 8.4. The manner in which two chains of $\alpha$-hemoglobin and two of $\beta$-hemoglobin form the complete molecule of hemoglobin. (Photograph by courtesy of Steiner, *Life Chemistry*, Van Nostrand Reinhold Company, New York.) Width of photograph = 12 nm.

I

FIG. 8.5. The hemoglobin molecule showing two heme groups end on. (Photograph by courtesy of Steiner, *Life Chemistry*, Van Nostrand Reinhold Company, New York.)

amount of space, just leaving wide enough gaps for the oxygen to enter, but also becoming thermodynamically stable by arranging all the polar side-groups ($-NH_3^+$, $-COO^-$, etc.) on the outside of the molecule (and hence they are solvated by the water in blood) and placing the hydrocarbon side-groups in the interior of the molecule. The heme groups sit in clefts in the hemoglobin, two being visible in Fig. 8.5. It is fascinating to note that when one oxygen molecule has become attached to a ferrous on a heme group the message is relayed to the other heme groups and then the second oxygen reacts much more quickly than the first, the third and fourth oxygens being progressively faster still. This 'message' is passed on via structural changes in the molecule; remember that the cleft clamps down and narrows when an oxygen is present and this causes the two $\beta$ chains to be moved relative to each other. Finally, the ferrous is not irreversibly oxidized up to ferric as it would be in aqueous solution because this involves forming an $O_2^-$ ion. However, the environment in the cleft is all non-polar (soft) and strongly discourages this increase in charges (HSAB). Hence the $Fe^{3+}$ and $O_2^-$ are very unstable if separated.

### 8.8 *The chemistry of vitamin $B_{12}$, a coenzyme* [14]

Vitamin $B_{12}$ was the first metallo-living system to be investigated in great depth. The work, done by Dr. R. J. P. Williams and his associates at Oxford, stands as an example, to those of us who study simpler systems, of the care and patience necessary to our studies and also as an example of how several techniques are needed to get an overall picture of these complex systems. The studies have already been reported with a clarity that cannot be improved upon but their approaches will be summarized here to illustrate the multi-technique approach.

The $B_{12}$ vitamins, some of the most complex non-linear polymeric natural products known, all contain cobalt (III?) as shown in Fig. 8.6. The metal is held inside a four coordinate corrin ring, assumed planar, and to two axial ligands, X and Y, above and below the plane of the ring. The cobalamin series have $Y =$ benzimidazole whereas the cobinamides have $Y = H_2O$. X may have many formulæ, among which is one giving a metal–carbon bond. Many cobalamin and cobinamide complexes were prepared for 20 different X ligands spanning oxygen, halide, nitrogen, sulphur and carbon donor atoms and these were investigated using visible and u.v. absorption spectra, i.r., c.d., n.m.r., e.p.r., and formation constant measurements.

8.8.1 *Absorption spectra*, recorded as X, was varied from hard donor $Cl^-$ through $Br^-$, $I^-$, $OCN^-$, $SCN^-$, $SeCN^-$, $C_2H^-$, $C_2H_3^-$ to soft donor $C_2H_5^-$, changed drastically due to electronic and conformational changes in the corrin ring (see Fig. 8.7). The temperature dependence of these spectra indicate there are two forms in equilibrium having an isosbestic point. One of these equilibrium forms is considered to be six coordinate as in Fig. 8.6, the other five coordinate, the Y = H₂O or benzimidazole molecule being entirely lost or very distant. Five coordinate cobalt

Cobinamide
R = NHCH₂CH (OH) CH₃

Cobalamin
R = NHCH₂CHMe

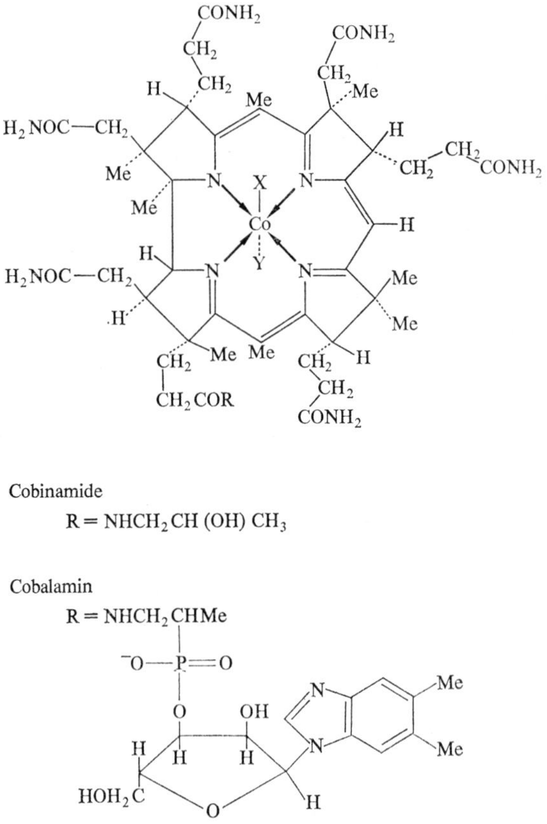

Fig. 8.6. The structure of cobalamins and cobinamides.

(III) is most unusual and is more often associated with cobalt (II) (low spin). N.m.r. evidence confirms this equilibrium theory and also indicates that the 5 ⇌ 6 coordinate exchange is rapid, most unlike the inert cobalt (III) complexes. Since the five coordinate cobalt (III) is the same form as the transition state for a ligand replacement in simpler cobalt (III) complexes, Vallee and Williams have called this equilibrium between the two states 'entatic'.†

† 'Entatic' is defined on p. 128.

*Circular dichroism* spectra parallel those of the change in absorption spectra as the donor power of the X ligand is altered. The c.d. spectra also indicate changes in the conformation of the complexes arising from the flexibility of the corrin ring.

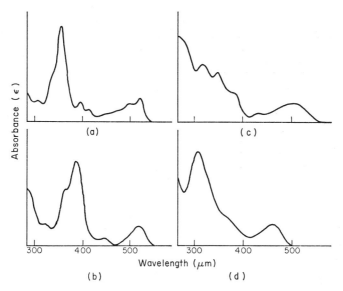

FIG. 8.7. Absorption spectra of different corrins as the electron donor power of X is increased from (a) through to (d). From Hill, Pratt and Williams [14].

8.8.2 *Infra-red spectra*   The *infra-red spectra* of a series of cyanide complexes in which $Y = CN^-$ and X is varied may be examined (from the point of view of the cyanide stretching frequency) to measure the trans effect of the X group. As X is changed from hard water to soft $C_2H_5^-$, the $CN^-$ is seen to be held less and less firmly. (The i.r. stretching frequency of $CN^-$ when $X = H_2O$ is 2130 cm$^{-1}$, when $X = C_2H_5^- = 2082$ cm$^{-1}$. Free $CN^-$ has a value of 2078 cm$^{-1}$.)

8.8.3 *Nuclear magnetic resonance spectra*   The nuclear magnetic resonance spectrum of vitamin $B_{12}$ is fortunately quite easy to correlate as the corrin ring has but one hydrogen bound to the unsaturated system. This is at $C_{10}$ and once again the spectral changes of this $C_{10}$ —H as ligands and as temperature are varied yield interesting results. The electron density on the ring is seen to increase systematically with the donor power of X. Again when we change from six coordinate to five coordinate the $\tau$ value changes too quickly to be explained by electron density changes alone and so a ring conformational change is postulated. The peaks conveniently group themselves into the high $\tau$ area, where hydrogens on the carbon ligands bound to the cobalt may be identified, and low $\tau$ area, where the benzimidazole hydrogens are observed. Very sensitive n.m.r. equipment can just identify parts of the $B_{12}$ molecule that are in contact with the enzyme.

8.8.4 *Equilibrium studies* Equilibrium studies in solution have reported the formation constants of a number of X and Y complexes. These indicate (a) that with soft X donors (such as $C_2H_5^-$) the Y group is very loosely held, i.e. the five coordinate species is possible, and (b) that $B_{12}$ has a Co(III) that falls into the soft classification of $[Co(CN)_5H_2O]^{2-}$ rather than the hard $[Co(NH_3)_5H_2O]^{3+}$ (cf. p. 46). Hence it is not surprising that $B_{12}$ has a strong affinity for sulphur and carbon ligands.

8.8.5 *Structural investigations* Structural investigations confirm that soft X donors, via the *trans* effect, cause exceptionally long Co—Y bonds. They also observe the buckling of the corrin ring that we have called conformational changes.

8.8.6 *E.p.r. measurements* These can investigate the first reduction product of $B_{12}$, $B_{12r}$—a tetragonal low spin cobalt (II) complex. The hyperfine structure is well resolved and e.p.r. promises to be an excellent probe of biological environments.

8.8.7 *Summary of the results of physical studies* The $B_{12}$ coenzymes undergo small changes when added to an enzyme and these can be greatly enhanced by adding substrates or inhibitors, i.e. changing the X and Y groups. One group may be held quite loosely by a very long bond and this is ideal for catalytic reactions as the cobalt lies close to its normal $S_{N1}$ mechanism transition state. The oxidation state of the cobalt may vary from III towards II depending on the nature of X and Y. Hence the activation energy for redox reactions is low. The corrin ring is flexible and can change shape to suit the 'environment of the day'. Cobalt–carbon bonds are possible especially if symbiotically encouraged by other soft ligands.

8.8.8 *Biological functions of $B_{12}$* Functions of $B_{12}$ include (a) the transfer of methyl groups and the rearrangement of carbon skeletons [Co–carbon bonds], and (b) the reduction of ribose [a cobalt redox reaction].

## REFERENCES

[1] TAYLOR, R. J., *The Physics of Chemical Structure* (Unilever Educ. Booklet, 1969).

[2] *Physical Methods in Advanced Inorganic Chemistry*, Ed. H. A. O. Hill and P. Day (Interscience, New York, 1968).

[3] SUTTON, D., *Electronic Spectra of Transition Metal Complexes* (McGraw-Hill, London, 1968).

[4] COTTON, F. A. and WILKINSON, G., *Advanced Inorganic Chemistry*, 2nd edn (Interscience, New York, 1966).

[5] MASON, S. F., *Chemy Brit.*, 245 (1965).

[6] BELLAMY, L. J., *Infra-red Spectra of Complex Molecules* and *Advances in Infra-red Group Frequencies* (Methuen, London, 1962 and 1968).

[7] NAKAMOTO, K., *Infra-red Spectra of Inorganic and Co-ordination Compounds* (Wiley, London, 1963).

[8] CARLSON, R. H. and BROWN, T. L., *Inorg. Chem.*, **5**, 268 (1966).

[9] CARRINGTON, A., *Chemy Brit.*, 301 (1968).

[10] COVINGTON, A. K., *Chemy Brit.*, 388 (1969).

[11] ÖSTERBERG, R. and SJÖBERG, B., *J. Biol. Chem.*, **243**, 3038 (1968).

[12] HARDING, M. M., *Chemy Brit.*, 548 (1968).

[13] KARTHA, G., *Accounts Chem. Res.*, 374 (1968).

[14] HILL, H. A. O., PRATT, J. M. and WILLIAMS, R. J. P., *Chemy Brit.*, 156 (1969).

*Chapter 9*

# CONCLUSIONS AND THE FUTURE

---

OUT OF the swamp of uncertainty concerning many aspects of metal ion chemistry *in vivo* there emerge certain areas of dry land where our facts are firmly established. These areas must be both extended outwards and have theories built upon them. These concluding pages will survey the extent of our knowledge about metalloenzyme systems, consider the usefulness or otherwise of studies on model systems and indicate the areas that are expected to develop in the future.

## 9.1 *Active centres in enzymes*†[1]

The impression is widespread that active centres are a relatively recent discovery and that the concepts of amino acids being in sequence in peptides and of peptides being combined into enzymes were known long before. In fact, active centres were suggested many years before the amino acid polymer ideas. Some forty years ago it was postulated that active centres consisted of amine, carboxylic and sulphur groupings that attracted substrates. However, only those molecules having the correct configuration could be attracted and held and of these only a proportion could be bent or configurationally altered so rendering them liable to undergo a chemical change, i.e. they become activated.

Nowadays the active centre is known to be that part of the enzyme that gives it its catalytic properties (see Fig. 9.1). The area is about 1·5–2·0 nm in diameter and is not always confined to one peptide chain as in the figure. The components of the active

† Active centre and active site are taken as synonymous.

centre have been identified by a variety of methods, for example, 1. by pK measurements with, and without, the substrate present, and 2. by degradation studies such as those performed on papain (an enzyme found in papaya fruit juices). Mercury renders papain

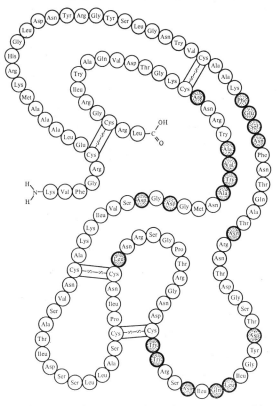

FIG. 9.1. A two-dimensional representation of the structure of lysozyme. The white amino acids are the ones that line the cleft in which the enzyme activity occurs. From Quastel [1]. The amino acid abbreviations are defined in Fig. 3.1 on p. 24. In addition, Asn = asparagine and Gln = glutamine.

inert, hence it must occupy and block off the active centre. Leucine aminopeptidiase is another enzyme that will degrade the mercury papain complex until only one-third of the original amino acids remains. If the mercury is then removed from this remainder, the latter is found still to have its catalytic activity. Hence we conclude that all 120 amino acids in papain were not really necessary to enzyme activity and were certainly not part of the active centre.

In all of the enzymes investigated to date, the centre has been found to be a crevice, not a protrusion. This (a) permits many more active groups to interact with the substrate, and (b) Perutz has suggested that the lining of the crevice in lysozyme is mainly composed of hydrocarbons and at the end of the crevice we have the polar groups necessary for activity. Hence the reaction path is assisted as the substrate enters the crevice through a corridor of low dielectric constant and is attracted only by its goal at the end of the crevice. Metal ions are frequently needed in these crevices, e.g. zinc is essential to carbonic anhydrase activity, there being one atom of zinc per molecule of protein. A thiol group such as one finds in cysteine, a protonated amine such as that in arginine or asparagine, and a tyrosine residue have all been identified as being in the vicinity of the metal [2], [3]. Zinc or cobalt can function in this enzyme. Copper, nickel, manganese, cadmium and mercury can also be bound but have little enzyme activity.

Allosteric effects have been observed in enzyme activity. Inhibition may occur at some distance removed from the active centre. The attachment of the inhibitor causes a change in the enzyme structure and so the geometry of the active site is distorted. This effect is commonly observed with 'feedback inhibition' in which the end product of a biological reaction exerts the inhibitory effect on the enzyme that produced it. This may be viewed as a kind of buffering of the catalyst.

Metalloenzymes are conveniently considered from the viewpoint of the transport and storage proteins and then the redox proteins.

## 9.2 *Transport and storage metalloproteins* [4]

These proteins, which involve metals, may be divided into two sub-divisions, (a) those which control metal ion concentrations (sequestering proteins) and (b) those which contain metals and control substrate (e.g. $O_2, CO_2$) concentrations. (a) *All* the metal ions in Chapter 2 are controlled in this manner, the iron and copper carrier proteins in the blood being the most well characterized. These proteins not only control the metal concentration and carry it to the site that requires it, but in addition, they can act as catalysts in placing the metal ion into its new molecule, e.g. transferrin transports iron to a porphyrin ring and then catalyses the insertion of the iron into this ring to make the heme group. If we search one step further back in this scheme we find that the transferrin concentration is controlled by another metalloenzyme (ceruloplasmin). This contains copper and controls the amount of iron (III) in serum and thence the amount of transferrin that can be formed. (b) Copper and iron carrier proteins are of three varieties—porphyrin–iron oxygen carriers (e.g. hemoglobin and myglobin), non-heme iron proteins (e.g. hemerythrin), and the copper proteins (e.g. hemocyanin). The geometry of the metal ion's bonds in these carriers is ideally suited to its role of addition of molecular oxygen and then its release (see Fig. 2.4), the iron (II) in hemoglobin and myglobin is held by five coordination positions so that the oxygen can travel to and from the sixth position. Had the iron (II) had its normal six coordinate geometry, the activation energy of adding oxygen would have been excessively high. Similarly, the copper in hemerythins and hemocyanins is entatically held so that it too has a vacant coordination position. However, with hemerythin and hemocyanin the oxygen attaches itself between adjacent positions on two neighbouring metal ions:

$$O_2 + [>Cu \quad Cu<] \rightarrow [>Cu \; O_2 \; Cu<]$$

The storage proteins merely hold the metal ion until a more powerful protein arrives and acquires the metal.

## 9.3 *Redox proteins involving metals*

This heading includes the metals in very many catalytic roles— electron transfer, oxygen atom and hydroxyl group incorporation, and hydrogen ion and atom removal. In addition, the processes are so arranged that the energy liberated at one site can be used at another. If an enzyme has two or more separate sites that activate a substrate or if the enzyme is really a system of smaller enzymes connected together and moving as a single unit, it is known as a many headed or a multi-enzyme. These more complex enzymes are ideal for conserving, and obtaining the maximum use of, chemical energy. It is this conservation and ultra efficiency as catalysts that distinguishes biological systems from laboratory models.

From Table 9.1, hydride transfer is seen to be associated with

TABLE 9.1.  A variety of metalloenzymes as listed in Table 1A of reference [7]

| Enzyme | Class of reaction | Metal |
|---|---|---|
| Carboxypeptidase | Peptide hydrolysis | Zn |
| Alkaline phosphatase | Phosphate ester hydrolysis | Zn |
| Carbonic anhydrase | Hydration | Zn |
| Alcohol dehydrogenase | Dehydrogenase (NAD) | Zn |
| Aldolase | Addition to carbonyl | Zn |
| Glycol dehydrase | Rearrangement | Co |
| Carboxytransphosphorylase | Phosphate transfer | Co |
| Phenol oxidases | Mixed function oxidase | Cu (Fe) |
| Cytochrome oxidase | Terminal oxidase | Cu, Fe |
| Pyruvate oxidase | Oxidation | Mn |
| Cytochrome c | Electron transfer | Fe |
| Ferredoxin | Electron transfer | Fe |
| Xanthine oxidase | Oxidation | Fe, Mo |

zinc, which does not have two different oxidation states in solution, but oxygen and electron transfer are often assisted by iron and copper and less frequently by manganese or molybdenum. (We have already seen how the iron and copper absorb oxygen reversibly.) For reversible electron transfer to the iron or copper a low activation energy necessitates both oxidation states having the same geometries. Failing this, the electron would like a compromise geometry between the two states, i.e. copper would be intermediate between tetragonal copper (II) and linear or tetrahedral copper (I). Iron prefers octahedral coordination in both oxidation states, although the bond lengths are different. It is not surprising, therefore, that e.p.r., Mossbauer and redox evidence shows the metal ion's bonds to be in an *entatic* state (from the Greek, meaning state of suspension or stretch) between the configurations of the two stable oxidation states. As the metal ion transfers the electron, it may be imagined as changing from one geometry of bonding upon receiving the electron into another upon releasing it.

From the conservation of energy angle, there are good reasons for combining several of these aforementioned functions (oxidation, electron transfer, hydride addition, etc.) into one enzyme system. These combinations fall into two categories—many headed enzymes and multi-enzymes.

## 9.4 *Many headed enzymes*

These are of high molecular weight (see Table 9.2) and may be

TABLE 9.2.  Some many headed metalloenzymes. (The number of metals per enzyme is shown in brackets.) From Table 6 of reference [4]

| Enzyme | Metals | Mol. wt. X $10^3$ | Function |
|---|---|---|---|
| Ferredoxins | Fe(2–7) | 5–10 | Removes H, or electron transfer |
| NADH-oxidases | Fe(2–6) | ? | Removes H |
| Flavin-oxidases (succinic, dehydroquinone) | Fe(4) flavin | ? | Removes H |
| Xanthine-oxidase | Fe(6) Mo(2)flavin | 250 | Oxidation of xanthine |
| Dihydro-orotic dehydrogenase | Fe(2) flavin | 62 | Removes H |
| Ascorbic acid oxidase | Cu(6) | 150 | Removal of H atoms |
| Phenol oxidases | Cu(2–8) | 30–300 | Removal of H atoms—melanins |
| Laccases | Cu(2–4) | 120 | Removal of H atoms |
| Monoamine oxidase | Cu(4) | 225 | Amine→aldehyde (pyridoxal coenzyme) |
| Plastocyanin | Cu(2) | 20 | Electron transfer |
| Ceruloplasmin | Cu(8) | 160 | Cu transport? Fe oxidation? |
| RHP-protein | Fe(2) | 30 | Bacterial respiration |

considered as small droplets of separate phases in water. They consist of several separable sub-units similar to the enzymes just described (these simple ones are often called single headed enzymes) but when combined in one whole unit, it is seen to have just one function. Many headed enzymes contain different metals, or many atoms of the same metal, either in the same, or different, oxidation states. In general, oxidation is usually by dehydrogenation rather than oxygen incorporation even though molecular oxygen is still the oxidizing agent, e.g. ascorbic acid oxidase has a first reaction product of the ascorbate free radical due to hydrogen atom abstraction. This is because this many headed enzyme absorbs the substrate at one site and the oxidizing agent ($O_2$) at another. If the reaction is described,

$$4H^+ + 4e + O_2 \rightarrow 2H_2O$$

the necessity of having the metal ions in the active site is clear as they are there to act as electron reserves or sinks.

## 9.5 *Multi-enzyme systems*

These are much larger particles of phases other than the medium water. They are mainly found as cell membranes where they are in a lipid-type matrix (refer to ion pumps on p. 10). They are capable of directing complete sets of complex reactions; the present day picture of these systems, and there is still a great deal more research needed, is:

1. Different ends of the multi-enzyme system may have very different redox potentials so that they can react with different substrates.

2. Electron transfer appears to be, not along protein chains, but by a hopping mechanism from one reactive site to another. The speed of this hopping appears to be under some sort of metabolic feedback control such as allosteric effects.

3. The chain of the active sites may contain energy traps, i.e. branches where energy not required immediately can be stored (see Fig. 9.2). Such a reversible process, called the mitochondrial

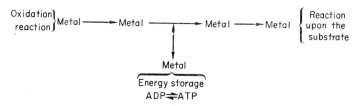

FIG. 9.2. Diagrammatic representation of energy traps in a many headed or multi-enzyme system.

respiratory chain, occurs in the inner membrane of cells.

Thus we can now see the constructional scheme of life:

atoms→amino acids→peptides→proteins→simple enzymes
→many headed enzymes→multi-enzymes→cells and life.

In this scheme (a) both simpler and the more advanced proteins have one common factor in that they often have an inorganic part and an active site, (b) our understanding of these systems is still so meagre that we cannot begin to model and mould enzymes for new functions—at least, not outside our bodies, inside our systems we are always adapting ourselves to our environment and (c) nature appears to have used common building bricks as far as possible—just 26 amino acids suffice for most constructional purposes. Thus, the whole process may be summarized as 'life is an organized, efficient polymer'. Clearly the skills and expertise of scientists from many fields are required to probe such systems.

## 9.6 *Working with model systems*

The two extremes of the attitudes to research are as follows. 1. The 'groundwork', 'preliminary investigations', 'model studies' approach which may be crudely described as trusting that experiments on sufficient simple systems will always give rise to 'fortunate accidents' (e.g. the discovery of penicillin) in which the new theory, drug or mechanism reveals itself. It is important to understand that expertise is required to recognize this discovery when it happens. 2. There is also the 'tackle the living system immediately' approach, jumping into models only if established theory appears inadequate or wrong when applied to *in vivo* reactions. (This was the main approach to vitamin $B_{12}$ in Chapter 8). With the proliferation of publications on model systems in different tongues from different viewpoints, approach 1 is becoming less and less likely to give the fortunate accident, if only because the efficiencies of our current awareness systems are decreasing under load, and so we are spending more and more time away from the laboratory, in handling the literature. In new drug design the shredding process that selects the one acceptable, from 10,000 possible, new drugs (see p. 6) is a tremendous chore. Now that more biochemical mechanisms are being discovered it is hoped that this task (which is sometimes nothing more than searching for the fortunate accident) can be reduced by more chemical planning going into the design of new therapeutics. Hence the route one chooses to approach a piece of research resolves itself into answering the scientific equivalent of the question: 'Ought we to pray for rain or to go out and dig irrigation schemes?' The author believes that a compromise approach is necessary but one that has a majority of direct attack.

Some generalizations are now listed concerning the uses of models in studying living systems.

1. Models are useful for establishing principles, e.g. (a) Copper ions prefer ligands with nitrogen donor groups rather than sulphur. (b) Copper usually forms stronger bonds than does zinc. (c) Wang, in considering the reversible attachment of oxygen to an iron (II), proved that this was possible in a non-polar environment by using absorption experiments of oxygen on to imidazole derivatives on a polystyrene film [5].

2. Very elaborate methods of analysis (e.g. e.p.r.) can, and ought to be, tried out first on model systems.

3. Model systems *cannot* be used to indicate the attracting power that an organism has for a metal ion, e.g. iron containing bacteria, Ustilago, grown in an iron deficient medium can emit a chelating agent (a substituted cyclic polypeptide $\beta \doteqdot 10^3$ 1 mol$^{-1}$) which extracts the iron from the stainless steel vats in which they are grown [6].

4. It ought to be remembered that (a) models are ordinary and life is extraordinary, and (b) laziness usually bids us use the simplest model. Thus, although models can establish 95% of how an enzyme reacts, we usually find also an unusual, unpredictable 5%. The moral to be learned here is that we ought to have an open mind to this question and not be tempted to set aside results as incorrect just because they do not fit the general, *in vitro* pattern, e.g. (i) five coordinate cobalt complexes are unlikely *in vitro* and yet are often encountered *in vivo*; (ii) we know of no ligand that prefers molybdenum to iron and yet xanthine oxidase appears to have sites that do just this.

5. Drugs: Sometimes *in vitro* tests of a new drug may prove positive whereas *in vivo* tests are negative either because the drug is digested or destroyed before reaching its required site or because it may be poisonous to the host. Similarly one finds vice versa, e.g. *in vitro* prontosil has no therapeutic activity in containing streptococcus infections whereas, *in vivo,* having been metabolized, it gives sulphanilamide whose effectiveness needs no further acclaim.

Our considerations of model systems fall into three areas:

(a) models involving metal ions;
(b) models involving water, and
(c) the use of animals as models.

9.6.1 *Models involving metal ions* [4]   Redox potential investigations on model systems have shown that if we distort the bonds on a metal ion suspended between two oxidation states, away from their usual configurations, we destabilize the higher valence state more than the lower and, as a consequence, the redox potential becomes more positive. Hence, when one finds, *in vivo*, that the redox potential of the oxygen transporter hemocyanin is high, it indicates that the copper is inclined to have a low oxidation state and irregular coordination. Further, if we then consider the heme iron proteins, they sometimes have much higher redox potentials than do their model systems. Spectral investigations reveal that the disfiguration causing this effect arises because one of the two groups that ought to be above and below the plane of the heme group is unable to come close enough to the iron for normal bond formation; in fact, only one of these two groups actually reach the heme iron. This gives heme proteins their high spins. If we, once more, invoke the use of model experiments and force a second perpendicular group to come towards the iron, monitoring the redox potential and the spins as we do so, one finds a maximum in the redox potential versus distance away curve that corresponds to the change over in spin states. In this example, model systems have supported *in vivo* suggestions.

'Models' may be obtained by metal ion substitution of metalloenzymes [7]. Metal substitution has the advantages that (i) it is milder than organic modifications, and (ii) it can be used to provide a unique electronic marker in the centre of enzyme active sites. The technique involves changing unexciting metal ions for metals having more suitable characteristics for detection. The more difficult metals to be studied are substituted by changing magnesium to manganese, zinc to cobalt, calcium to a rare earth, and potassium to thallium. We might substitute the enzyme with a metal ion that suits the technique chosen, e.g. for e.p.r. and n.m.r. use manganese (II) or iron (III), for redox studies use copper (II) or iron (III), or we may make several models from a graded series

K

of metals, e.g. traversing the extended Irving–Williams series. However, model systems do not necessarily parallel the *in vivo* situation (see Fig. 9.3). Clearly, as Lewis acid strength increases

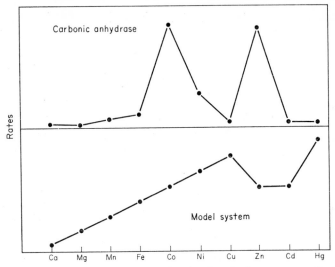

FIG. 9.3. The rate of enzyme reaction of carbonic anhydrase compared with an idealized model system. From Vallee and Williams [7].

from calcium to mercury, there is no parallel trend in carbonic anhydrase activity. Thus the complexity of enzymes as catalysts cannot always be directly related to the strength of the bonds involved and be readily mimicked using models. Instead of changing the metal ions one can, of course, study models by changing the substrates and many experiments along these lines were made concerning the cobalt–carbon bonds in vitamin $B_{12}$ [8].

Model systems have been used for studying the selective cleavage by alkali when in the presence of transition metal ions. Freeman *et al.* have found that the amide bonds in simple peptides which are usually easily hydrolysed to give amino acids again are protected by transition metal ions, the best protectors being copper (II) and nickel (II), e.g. consider a hexapeptide of glycine being hydrolysed.

$$\overset{1}{Gly} - \overset{2}{Gly} - \overset{3}{Gly} - \overset{4}{Gly} - \overset{5}{Gly} - \overset{6}{Gly} \overset{OH^-}{\rightarrow} (Gly)_5 + Gly \qquad (1)$$

$$(Gly)_5 \overset{OH^-}{\rightarrow} (Gly)_4 + Gly \qquad (2)$$

If the reaction is performed without metal ions *complete* hydrolysis occurs, and no hexa-, penta- or tetrapeptide is detectable after 15 min. However, when transition metal ions are used along with the alkali, they selectively inhibit hydrolysis so that step (1) is complete after a few minutes, step (2) after 15 min and glycyl-glycine is present only after 30 min. Also, by synthesizing the hexapeptide with a $^{14}C$ in the terminal —COOH, it was possible to show that selective cleavage occurs in step (1) between residues 5 and 6. It is believed that the copper (II) complexes involved have coordination by four nitrogen donors so that four or five residues of a peptide are protected against hydrolysis by the presence of the metal ion. This new principle, discovered on model systems, has obvious applications to degradative studies of sequence determination of amino acid residues in the peptides and proteins found *in vivo*.

9.6.2 *Models involving water*    The role of water in chemical equilibria is mainly one of governing the entropy term, $\Delta S$; e.g. the presence of water favours the removal of a proton from an —$NH_3^+$ compared to a —COOH since it requires more energy to produce and separate two *charged* particles and to tie up two aliquots of water of hydration (—$NH_2 + H_{aq}^+$ c.f. —$COO_{aq}^- + H_{aq}^+$). Hence there is a smaller entropy contribution to $\Delta G$. This may be seen from the pKs in water and is a very much reduced concentration of water— 95% methanol/water [9].

$$pK_{-NH_3^+}^{methanol/water} - pK_{-NH_3^+}^{water} = 0.3$$

$$pK_{-COOH}^{methanol/water} - pK_{-COOH}^{water} = 2.3$$

Clearly solvation has a much greater influence upon groups that produce two charged ions (—COOH) than it does on groups producing only one ion (—$NH_3^+$ or imidazole —$H^+$).

We might generalize and state that bonds that may be ionized in our bodies are very water dependent. To ensure ready ionization and ion separation requires as much water as possible. Conversely, to prevent ionization, we require a scarcity of water (as in the $Fe^{2+}$—$O_2$ case in oxyhemoglobin). The relative hydrophobia or hydrophilia of species may be used as a simple screening test for anaesthetics and stimulants—the dough test. First, let it be said that the chemical mechanism of anaesthesia, or the converse operation of stimulation, is not yet fully understood. Nevertheless, it is known that general anaesthetics, e.g. diethyl ether, form hydrates and therefore they compete with the brain cells for the available water. If they can rob cells of water, they can also rob ordinary household baking dough, and this they do: anaesthetics make dough tougher, as though it contained less water. On the other hand stimulants, e.g. caffeine or strychnine, which have the opposite effects to anaesthetics on the central nervous system, have the opposite effect on dough—they make it crumbly, not tougher. Further, theobromine, which is chemically very similar to caffeine but without its stimulant properties, does not crumble dough (see Fig. 9.4). The search for new anaesthetics is very much a hit-and-

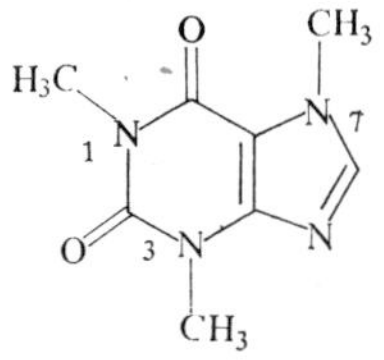

Caffeine
(1,3,7 - Trimethylxanthine)

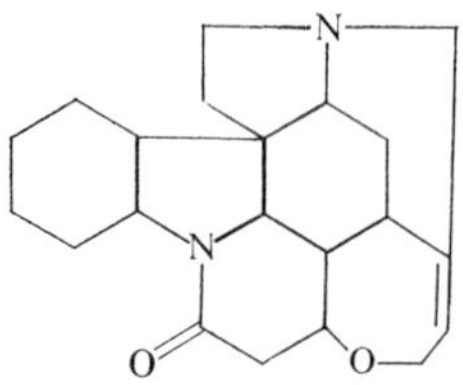

Strychnine

FIG. 9.4. Two stimulants, caffeine and strychnine. Theobromine has a H where caffeine has a $CH_3$ in the 1 position.

miss affair depending on the screening of a huge number of chemicals. At present, the only screening tests for anaesthetics are animal tests and a piece of dough is much less expensive than even the humblest laboratory mouse. This search for new anaesthetics and for analgesics (pain killers) is a serious matter. Many analgesics in use today have undesirable effects upon humans, e.g. morphine (Fig. 9.5) soon renders the body physically dependent upon continued doses [10]. It also gives respiratory depression and in obstetrics can give foetal distress. Other after-effects of morphine include fits of gloominess and depression but these effects can be removed by administering nalorphine, a specific narcotic antagonist. Nalorphine is a derivative of morphine but because it is used as an antagonist it would not be expected to have analgesic properties. In experimental animals it is not an analgesic but in humans it is as potent as morphine. Also it is non-addictive although it does give hallucinogenic after-effects.

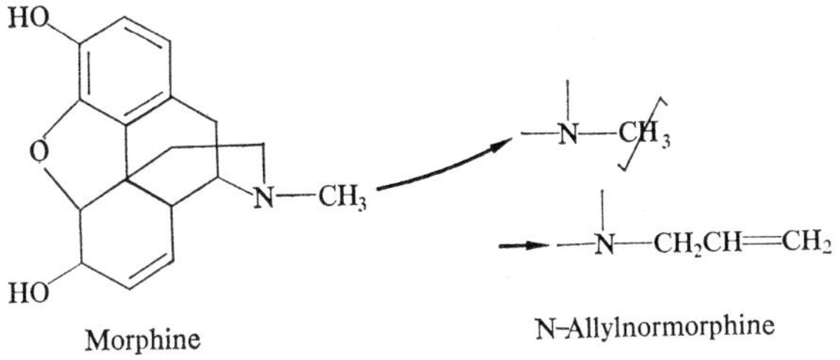

Morphine                          N-Allylnormorphine

FIG. 9.5. Morphine and nalorphine.

A lack of solvation results in precipitation and this is what happens in cataracts, gall stones and atherosclerosis. A cataract is a precipitation of calcium salts over the cornea of the eye; when a sufficient thickness has accumulated the eye can no longer receive incident light and blindness ensues. Treatments include either sequestering the calcium by bathing the eye in E.D.T.A. solution, or allowing the precipitate to build up into a hard crust and then lifting this crust away by a minor operation. Gall stones and atherosclerosis (hardening of the arteries that lead to the heart) involve both calcium and cholesterol. Gall stones are precipitates of calcium cholesterate and may be removed by elution or by surgery. Atherosclerosis is a more complicated disease that occurs in the middle layer of arterial walls, called the intima [11]. The intima becomes thickened by an increase in the number of smooth muscle cells and by the deposition of cholesterol (and its esters) and phospholipids and glycerides. This then encourages the precipitation of calcium salts (e.g. aragonite structured calcium carbonate has been identified) which cause the hardness and brittleness associated with atherosclerosis. This swelling and hardening of the wall results in a constricted artery giving an elevated blood pressure and encourages the coagulation of blood platelets and clotting (a thrombus). Finally, there is a possibility of an aneurysm of this blocked, brittle vessel. Here, there is a direct challenge to

solution chemists to devise means of redissolving bodily precipitates and, in the longer term, of dietetically preventing them from
ever recurring.

9.6.3 *The use of animals as models*   The advantages of the use of
animals are that it conserves chemicals, saves human lives and
also, as many animals have much shorter lives and gestation
periods than humans, several generations may be studied in a
comparatively short time. The disadvantages are that some of
their chemistries differ from that of humans (e.g. insulin, see
p. 28) and a drug showing a positive test in animals may show
negative in man. Also, the law forbids the use of any experiment
that causes animals pain whereas humans can volunteer for these
experiments. Finally, animals are no good for behavioural studies.
The closest we can mimic human behaviour in animals is by
experimenting on a specially bred strain of pigs, called the Hanford
miniatures. These have similar anatomy and physiology to humans,
have a tendency towards fatness, and they develop stomach
ulcers and cardiovascular diseases resembling man. However,
because of animal difficulties, it is better to employ chemical
screening before screening new drugs through animals.

## 9.7 *The future*

The future will only be bright if we can delete our present
ignorance of the connections between chemical structure and
biological activity. In this way alone can researches for new
drugs be planned; perhaps there will be a phosphorus era as there
was a sulphonamide era. 'New drugs' does not mean close ana-
logues to those already in use and having no obvious advantages
but rather reagents especially designed to minimize side effects (the
embolisms associated with hormonal contraceptive pills, the
drowsiness with antihistamines and the withdrawal problems
with morphine). Another aim must be to design more powerful
drugs that can function at lower concentrations.

Computerized medication cannot be many years away. HALTA-
FALL type calculations will soon be able to calculate the effect
upon every part of the body of swallowing 1, 2 or 3 capsules of
sedative or tablets of aspirin. In this way optimum doses may
be chosen so that palliative or curative effects can be balanced

against undesired side-effects. In the past, mixing two or more drugs has been frowned upon but this can now be turned to useful purpose as, by careful selection of the pairs, synergic effects may be exploited (this is the technique of having two well matched chemicals that together work more effectively than the sum of the two individually). Continuous theatre monitoring to give therapeutic computing for continuous intravenous administration and surgeries and clinics land-linked to computers will be a common occurrence before the end of the century. The 'inhibitor' in this process is not the limitations, or lack, of computers or the shortage of personnel searching for new drugs. The deficiency lies in the lack of chemical theories that are required to program computers and to steer the researchers. All too often speculation outstrips the precision and quantity of the data available.

If we now turn from the general to the particular, metal ions have been seen to be capable of directing the hydrolysis of peptides. Conversely, more researches will illuminate to what extent they regulate the *correct* synthesis of amino acids into useful peptides instead of the *incorrect* or *mad* synthesis into cancer peptides or toxins. For it has been shown that the fluid surrounding the growth of a cancer contains this toxin (a polypeptide of eight different amino acids, MW = 1900). This toxin prevents the production of normal cells by interfering with a specific stage (probably the *S* stage in which the hereditary material DNA is manufactured) in the life cycle of a normal cell. *In vitro* the polypeptide appears to prevent the nucleosides forming nucleotides. However, this toxic effect may be neutralized by other nucleosides such as deoxycytidine or thymidine. Malignant tumours are rich in these latter and so are not affected by the toxin. Thus normal cells are prevented from multiplying correctly, malignant cells are permitted to reproduce. Could a metal or its complex be chosen to destroy or put to good purpose this toxin? Of course, this would not be a cure for cancer but, at least, it would stave off the metabolic death that is so often associated with it.

Another peptide of current interest is the hormone calcitonin [12] (see Fig. 9.6) that helps to keep bones healthy rather than the weakness and brittleness of bone often found in the aged. Calcitonin controls the rate of breakdown of cells in the bone marrow

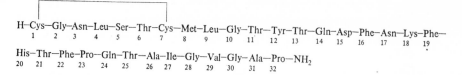

H–Cys—Gly–Asn–Leu–Ser–Thr–Cys—Met–Leu–Gly–Thr–Tyr–Thr–Gln–Asp–Phe–Asn–Lys–Phe–
  1    2    3    4    5    6    7    8    9   10   11   12   13   14   15   16   17   18   19

His–Thr—Phe–Pro–Gln–Thr–Ala–Ile–Gly–Val–Gly–Ala–Pro–NH$_2$
20   21   22   23   24   25   26   27   28   29   30   31   32

FIG. 9.6. The structure of human calcitonin M. Calcitonin D is the anti-parallel dimer of calcitonin M.

and hence the rate of release of calcium into the blood stream. Its administration leads to the speedier healing of fractures and it is used in the treatment of osteoporosis and bone cancer. It is surprising to find that human calcitonin differs by 18 amino acids from that of pig calcitonin—there being only 32 amino acids in the whole molecule. Ciba can now synthesize this hormone at reasonable cost. Clearly a great deal of research into metal–calcitonin complexes is necessary before it can be released as a general medicine.

Everyone must age and life is finite. Nevertheless, the pattern of ageing and the quality of life might well be metal ion controllable.

# REFERENCES

[1] QUASTEL, J. H., *M. & B. Lab. Bull.*, **8**, 18 (1968).

[2] WILLIAMS, A., *Q. Rev.*, **23**, 1 (1969).

[3] DOONAN, S., *R.I.C. Rev.*, **2**, 117 (1969).

[4] WILLIAMS, R. J. P., *R.I.C. Rev.*, 13 (1968).

[5] PERUTZ, M. F., *Chemy Brit.*, 14 (1965).

[6] SPIRO, T. G. and SALTMAN, P., *Structure & Bonding*, **6**, 116 (1969).

[7] VALLEE, B. L. and WILLIAMS, R. J. P., *Chemy Brit.*, 397 (1968).

[8] HILL, H. A. O., PRATT, J. M. and WILLIAMS, R. J. P., *Chemy Brit.*, 156 (1969).

[9] GRENTHE, I. and WILLIAMS, D. R., *Acta. chem. scand.*, 341 (1967).

[10] BENTLEY, K. W., *Endeavour*, 97 (1964).

[11] FLOREY, H., *Endeavour*, 107 (1967).

[12] RITTEL, W., *et al.*, *Helv. chem. Acta.*, **51**, 1738, 1900, 2057, 2061 (1968); *Chemy Brit.*, 94 (1969).

# GLOSSARY

IT IS difficult to write any book using a vocabulary and terminology that every reader will understand. In this interdisciplinary book such an aim is clearly impossible unless one labours every term encountered so that one reduces the level of sophistication to a shallowness that insults the reader. To overcome this barrier, a glossary of the more technical terms used is presented. Depending upon which discipline the reader has been trained in, half the terms ought to be everyday language, the remainder new ground.

Definitions and descriptions have been extracted from many of the references cited in Chapters 1–9 and from the *Shorter Oxford English Dictionary* (Clarendon Press, Oxford). However, the descriptions have been specifically written for the context in which the terms appear in this book. The page numbers where these terms appear will be found in the index.

Absorption band. When a molecule vibrates, in theory, it ought to absorb the exact infra-red energy ($E$) needed to cause the vibration $E=h\nu$, where $h$ is Planck's constant. Hence one expects an infra-red absorption peak at frequency $\nu$. However, *all* vibrations and deformations are somewhat interconnected and so the energy level step for the particular system under consideration may not be exactly the same each time the molecule vibrates. Thus a band of absorptions spread about $\nu$, rather than a sharp peak, is observed.

Accumulator. An apparatus for storing electricity.

Active centre. The portion of an enzyme that produces its catalytic properties, i.e. that is responsible for lowering the free energy of activation for the reaction being catalysed.

Active site. Synonymous with active centre.

Activity. From Table 6.1, activity = fugacity × concentration. Many thermodynamic relationships only apply to the concentration of species in ideal solutions (usually very dilute solutions). In more concentrated solutions the actual quantity of chemical that is truly available for reaction is called the activity. Hence thermodynamic relationships can be applied to non-ideal solutions as long as activities are used instead of concentrations.

ADP–ATP system. Adenosine-5′-triphosphate (ATP) is a carrier of chemical energy in a readily available form. When energy is required, ATP is converted to the diphosphate + energy

$$\text{ATP} + \text{acceptor} \rightarrow \text{acceptor} \sim \left( \begin{array}{c} O \\ \parallel \\ O{-}P{-}O \\ | \\ O^- \end{array} \right) + \text{ADP} + \text{energy}$$

Adiabatic calorimeter. An instrument for measuring quantities of heat that through its adiabatic properties ensures that no heat enters or leaves the reaction vessel.

Adult. A mature human being.

Ailment. Affliction, illness or suffering.

Albumins. These are a large group of soluble simple proteins that are present in eggwhite, blood and milk. Heat coagulates albumins. They can often be obtained as crystals.

Algae. A division of cryptogamic plants, including sea-weeds.

Allosteric effects. A change in the catalytic function of enzymes by small molecules (e.g. activators or inhibitors) being attached at a site remote from the active site. They are believed to result from structural transitions.

α and β chains in hemoglobin. Adult hemoglobin is constructed of two pairs of peptides (two called α and two β). The four chains fit together into a compact unit (see Figs. 8.4 and 8.5).

Alveoli. Cavities, chambers or pockets.

Anaemia. Lack of hemoglobin, blood or of red corpuscles in the blood.

Anaesthesia. Insensibility, loss of the feeling of sensation.

Analgesic. A pain killing chemical.

Analogue. Of similar attributes or uses.

Aneurysm. A blood containing tumour connecting directly with the lumen of an artery or formed by a circumscribed enlargement of an artery.

Anion. Faraday's term for an ion carrying a negative charge of electricity.

Antagonist. One which contests with another.

Antibiotic. Chemicals produced by bacteria, moulds, yeast or actinomycetes from their supporting media. They are capable of inhibiting the growth of other micro-organisms. Antibiotics are of two classes; bacteriostatic (halting growth and multiplication of bacteria) and bactericidal (destroying bacteria).

Antienzymes. These are very important compounds that arrest enzyme activity and so prevent digestive enzymes from digesting stomach walls, etc. They are always being manufactured during life but after death they are not made so the body begins to digest itself.

Antiseptic. Counteracting putrefaction.

Antizygotic mechanism. A mechanism for destroying a germ cell that has arisen from the union of two reproductive cells or gametes.

Apoenzymes. Enzymes minus their activators.

Arthritis. A disease causing inflammation of the joints.

'-ase' ending for enzymes. This ending is added on to a prefix closely related to the substrate, e,g. maltase catalyses maltose hydrolysis to glucose, oxidases oxidize hydrogen atoms into water. Not all enzymes end in -ase; some of the first to be discovered end in '-in', e.g. pepsin.

Atherosclerosis. A deposition of excess cholesterol and salts in the walls of blood vessels. This results in a thickening and eventual brittleness of the vessel walls. Further details are given in Section 9.6.2.

Bacteria. Microscopic cellular organisms found in (decomposing) animal and vegetable liquids.

Bantus. A tribe of original inhabitants of South Africa.

Bidentate. Some ligands are attached to a metal ion via *two* donor atoms. Under these circumstances the ligand is said to be bidentate.

Binding groups of peptides. In Chapter 4 this phrase means the donor atoms of peptides that are rich in electrons that may be donated towards metal ions.

Biochemistry. Pertaining to the chemistry of life.

Blood. The red liquid circulating in the arteries and veins of man and of higher animals.

Buffer. A system for curtailing the force of a forward reaction.

Ceruloplasmin. A copper containing glycoprotein that accounts for most of the copper in plasma. In Wilson's disease, the ceruloplasmin level is reduced.

Calcitonin. A hormone that controls the breakdown of cells in the bone marrow (see Section 9.7). It controls the rate of calcium release into the blood stream.

Calorimeter. An instrument for measuring quantities of heat.

Cancer. A malignant growth or tumour that has tendencies to spread and produce.

Carcinogen. An agent which induces cancer.

Cardiac. Pertaining to the heart.

Casein. A protein that is one of the chief components of milk. It is coagulated by acids.

Catalyst. A substance that produces or facilitates a chemical reaction but which, itself, undergoes no permanent change. A catalyst reduces the energy barrier (free energy of activation) for the reaction under consideration.

Cathartic. A medicine producing the second stage of purgation, of which laxative is the first and drastic the third.

Cation. Faraday's name for an ion carrying a positive charge of electricity.

Cell. An enclosed sac in organized bodies. The ultimate element in organic structures. It is a minute portion of protoplasm enclosed in a membrane.

Cerebrocuprein. An enzyme containing two copper ions and used for oxygen storage and transport in the brain.

Charge transfer. In Table 1.1 this refers to the movement of charges as ions from one part of the body to another.

Chelate ring. A ligand that is attached to a metal ion through two or more donor atoms is said to form a chelate ring. It originates from the Greek word for a crab's claw.

Chemical process. A related system of many chemical reactions.

Chemistry. The portion of science that deals with elementary substances, forms of matter and the laws governing their combinations and reactions when exposed to diverse physical conditions.

Chemotherapy. Medical treatment of disease using chemicals.

Childcarriage. Pregnancy.

Chlorophyll. The colouring matter of the leaves of plants. It is usually found inside cells as minute granules.

Cleavage of peptides by alkali. Fission or splitting asunder of peptides into individual amino acids.

Coenzyme. A chemical that activates enzymes. Unlike enzymes, coenzymes are not deactivated by heat. Many coenzymes are vitamins.

Cofactor. The extra molecule or ion that is necessary before an enzyme can function. There are 3 types; metal ions, coenzymes and primers (coenzymes that finish up complexed with the product of the enzyme reaction).

Colostrum. The first milk secreted by a mammal after parturition.

Complex. A group of constituents joined together.

Computer. A machine, usually electronic, for performing many calculations and obeying many instructions very rapidly.

Conception. The action of producing an embryo.

Conformation. The overall form that depends upon the arrangement of the constituent parts.

Contraceptive. A method of preventing uterine conception.

Convulsions. Irregular contractions or spasms of the muscles.

Coordinate bond geometry. A dative covalent bond is directional and the arrangement of these bonds around a metal ion can be called the coordinate bond geometry of the metal ion, e.g. tetrahedral or octahedral.

Corpus luteum. In the female, an ovary follicle sheds an egg and then the wall of this used follicle develops into the corpus luteum (yellow body) which proceeds to secrete estrogens (mainly progesterone). In a cycle that is not interrupted by pregnancy, the corpus luteum persists for 13–14 days and then degenerates and menstruation follows immediately. The gland is creamy grey in humans, not yellow. During pregnancy, the human corpus luteum is active for the first two to three months then the placenta takes over.

Countercurrent distribution apparatus. This is an instrument for separating two closely related chemical components. The process could be very time consuming without such an apparatus.

Crystal field stabilization energies. In a tetrahedral or octahedral complexed metal ion, not all the $d$ energy levels are equivalent although they were all equal in the free ion. Hence a lower energy state may be reached by placing electrons into the lower energy orbitals. The total energy thus gained is known as the crystal field stabilization energy. However, in order to gain this energy, energy sometimes has to be expended to pair off electrons of parallel spins (see Fig. G.1).

Cure. Medical treatment that successfully restores health; a remedy.

D configuration. The three-dimensional arrangement of groups around an asymmetric carbon atom. D amino acids are arranged

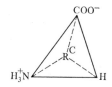

L are the mirror images of D.

Degradation studies. Investigations into polymers by chemically breaking them into their smaller constituent units, e.g. the amino acid sequence of a peptide chain is determined by successively removing amino acids from this chain and identifying each unit in turn.

DNA. Deoxyribonucleic acid. DNAs hold the codes which determine the order in which amino acids appear in proteins. Ribonucleic acids carry the information about these codes from the DNA to the sites of protein formation.

Dermatitis. A skin condition of inflammation.

Diabetics. Patients suffering from a pancreas that does not produce sufficient insulin so not all ingested glucose is metabolized and excesses appear in the blood and urine.

A maximum of two electrons may be found in any $d$ orbital. In the free metal ion all $d$ orbitals are of equivalent energy and may be depicted:

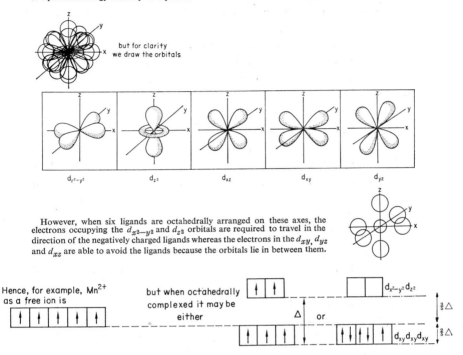

but for clarity
we draw the orbitals

$d_{x^2-y^2}$     $d_{z^2}$     $d_{xz}$     $d_{xy}$     $d_{yz}$

However, when six ligands are octahedrally arranged on these axes, the electrons occupying the $d_{x^2-y^2}$ and $d_{z^2}$ orbitals are required to travel in the direction of the negatively charged ligands whereas the electrons in the $d_{xy}$, $d_{yz}$ and $d_{xz}$ are able to avoid the ligands because the orbitals lie in between them.

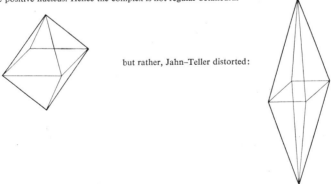

Hence, for example, $Mn^{2+}$ as a free ion is

but when octahedrally complexed it may be either

$\Delta$ or

$d_{x^2-y^2}d_{z^2}$

$d_{xy}d_{xy}d_{xy}$

$\frac{3}{5}\triangle$

$\frac{2}{5}\triangle$

Thus the *crystal field stabilization energy* (c.f.s.e.) $= +3\times\frac{2}{5}\triangle - 2\times\frac{3}{5}\triangle = 0$ or $5\times\frac{2}{5}\triangle$
— energy required to pair off electrons

*Jahn–Teller distortion*
    The distances between the ligands and the nucleus is dictated by charge considerations, e.g. the negative ligands are attracted to the positive nucleus but are held at a fixed distance away by the negatively charged electrons on the axes. However, if the electron distribution is not symmetrical, e.g. in a Cu(II) complex, the electron distribution is $d^2_{xy}$, $d^2_{yz}$, $d^2_{xz}$, $d^2_{z^2}$ and $d^1_{x^2-y^2}$: there is only one electron to keep the four ligands on the $x$ and $y$ axis at bay and consequently they are attracted in more closely to the positive nucleus. Hence the complex is not regular octahedral

but rather, Jahn–Teller distorted:

FIG. G.1. The crystal field theory and Jahn–Teller effect.

Diet. The pattern of eating. The food in daily use.

Dietetics. The study of the kind and quantity of food eaten.

Diffraction grating. A grating that can be used to resolve a beam of incident energy into its constituent wavelengths.

Disease. The condition of having a part of the body disturbed or deranged. A departure from health; an illness.

Diurnal variation. Daily changes or differences. The pattern is usually repeated every twenty-four hours.

Donnan membrane equilibrium. This is the equilibrium set up when a solution consisting of an electrolyte of two membrane diffusible ions is separated by a membrane from another solution containing a salt of a non-diffusible ion and then the system is allowed to equilibriate. The resultant distribution of the electrolyte will be unequal on both sides of the membrane. Plants use this method for obtaining their nutrients via their root systems (membranes).

Donor group. A chemical group having an atom that possesses spare electrons that may be donated towards a cation.

Drug. A medicinal substance.

Edema. When the osmotic pressure of the blood drops excess fluid accumulates in the intercellular spaces of connective tissue producing the condition known as edema (or oedema).

Electrode. The means (often wires or plates) of leading a current into an electrolytic solution so that it might produce ions. Or, a device that responds to the presence of ions in solution by producing a current or voltage, e.g. a calomel electrode gives a voltage response to $a_{Cl^-}$ in solution.

Electrolyte. A conductor of electric current. In solution this usually means a salt that ionizes and the ions may be separated by an applied voltage.

Electronic heavy atom marker. In X-ray crystallography, the intensities of the diffracted beams are dependent upon the electron densities of the unit cells. In complex crystals resolution is difficult unless some reference point can be selected. A metal ion in such a crystal has a very high density of electrons and so is ideal as a reference marker.

Electron microscope. A microscope-like device whereby objects are viewed not under visible light but under illumination from an electron beam (i.e. a much shorter wavelength). By this means, particles of molecular dimensions can be observed.

Electrostatic. Stationary electric charges.

Embolism. The occlusion of a blood vessel by a blocking agent such as a blood clot.

Emesis. Vomiting.

e.m.f. Electromotive force; electrical potential.

Enantiomorphs. A pair of crystals which are optical isomers and are non-superimposable mirror images of each other.

Entatic. Strained (configuration of bonds), e.g. they may be forced to lie away from their usual octahedral arrangements. The word is derived from the Greek meaning 'in state of suspension or stretch'.

Enthalpy. The heat content of a system. A reaction heat arises from an enthalpy change, i.e. the difference in the enthalpy contents of products minus reactants.

Entropy. The degree of disorder, randomness or freedom of a system. Any restriction placed upon the system (e.g. a strain in a chelate ring, a large amount of water of hydration) reduces the entropy.

Enzyme. A complex substance that causes, or controls the rate of, chemical transformations in materials in plants and animals.

*Escherichia coli.* A 'simple' intestinal bacteria containing approximately 3000 genes.

Estrogen = Oestrogen.

Excrete. To separate and expel from the system.

Feedback inhibition. Retroinhibition. The checking, buffering or controlling of a reaction by the end product of a biochemical sequence inhibiting one of the initial steps of the sequence.

Fibrinogen. A major protein present in blood plasma. It is required for the clotting of the blood. During clotting, fibrinogen is denatured into fibrin. Calcium ions and the enzyme thrombin are also essential for clotting.

Filtration. To pass a solution through a porous membrane so that larger particles are removed.

'Fingerprinting'. In Chapter 8 this refers to infra-red spectra. Just as every human has a unique set of fingerprints, so too,

L

every molecule has a unique infra-red spectra. Libraries of infra-red spectra can now be used to identify molecules.

Fluoridation. The addition of fluoride to water supplies so that the minimum level of 1 p.p.m. is maintained. The need for fluoride is described in Section 3.7.

Foetus. The young of viviparous animals inside the womb.

Fuel cell. An apparatus for storing energy as chemical energy and capable of giving back this energy when required.

Fugacity. The ratio of chemical activity to concentration; a constant to express the non-ideality of the solution. Definition is given in Table 6.1.

Gall bladder. The vessel in animals that contains the gall or bile.

Gastrointestinal tract. The stomach and intestine through which food is digested and absorbed.

Globulins. These are described in Section 3.4.4.

Goitre. An enlargement of the thyroid gland of the neck.

Gonorrhoea. An inflammatory discharge of mucous from the membrane of the urethra or vagina.

Gout. Painful inflammation of the smaller joints of the body.

Gramicidin S. A cyclic peptide that is used as an antibiotic. Its structure is unusual because it contains D-phenylalanine and also ornithine.

Heat sink. An object having such a high heat capacity that one can abstract heat from it, or add heat to it, without appreciably affecting its temperature, e.g. a large thermostat bath.

Hemichelate. An unusual chelate ring, e.g. in Section 4.2.1 there is one involving water. Literally half way between a chelate ring and a single coordinate bond.

Hemoglobin. The colouring matter of the red corpuscles of blood, which serves to convey oxygen to the tissues in the circulation.

Heterocyclic ring. A ring of at least two different types of atoms,

e.g. pyridine, , is carbon and nitrogen.

High spin state of $d^5$ manganese (II). Because of crystal field stabilization energy, octahedrally complexed manganese (II) may have its $d$ electrons arranged ⟨↑↓|↑↓|↑| | |⟩ rather than ⟨↑|↑|↑|↑|↑⟩. These two possibilities are called low and high spin, respectively. (See Fig. G.1.)

Hormone. A substance formed in an organ and serving to excite some vital process. Further details are given in Chapter 3.

Hydration shell. Ions in aqueous solution acquire two groups of water molecules. Nearest to the ion there is the primary hydration shell which is often considered to be water molecules acting as ligands to a metal ion. Outside this there exists a less firmly held layer of water molecules called the secondary hydration shell. Both anions and cations have these two shells of aquation.

Hydrogen bonds. Partial bonds formed by electrostatic attraction between a hydrogen atom and an electronegative atom on the same molecule or on a separate molecule. The range of bond strengths encountered are the order of $1-40$ kJ mol$^{-1}$. Hydrogen bonds are usually denoted by dots, e.g. A—H . . . B

Hydrolysis. The reaction of water with a metal ion to give a hydroxyproduct and a proton.

$$M^{n+} + H_2O \rightleftharpoons MOH^{(n-1)+} + H^+$$

Hydrophilic groups. Groups that are attracted to water and prefer to become hydrated, e.g. $COO^-$, —$NH^+$. Hard groups in the HSAB sense. Literally it means water loving.

Hydrophobia. The opposite of hydrophilia. Water hating. Soft HSAB groups,

e.g. —⟨◯⟩, —$CH_3$.

Hypnotic. A soporific chemical inducing sleep.

Hypothesis. A proposition or principle stated as a basis for reasoning or argument.

Hypothyroidism. An underactivity of the thyroid gland leading to sluggishness, increases in weight, loss of appetite, slowing down of the heart beat and a reduced metabolic rate in general.

Impurities. Dirt or foreign matter.

Infant. A child.

Influenza. A contagious zymotic disorder producing prostration and catarrh.

Infra-red spectra. When a molecule is bombarded with infra-red energy its atoms and bonds vibrate, twist and resonate. The energy to achieve these movements must be absorbed and a plot of infra-red absorption energy versus wavelength is called an i.r. spectra.

Ingestion. The action of taking in nourishment, usually through the mouth.

Inhibitors. A chemical that forbids or hinders a reaction.

Inorganic chemistry. The chemistry of elements and compounds, with the usual exception of compounds mainly of carbon and hydrogen. The correlation of theories of the elements and their compounds with experimental facts.

Insulin. A drug extracted from the islets of Langerhans in the pancreas and used in the treatment of diabetes.

Inter. Preposition meaning between, among, or amongst other things.

Intra. Prefix meaning on the inside, within.

Intravenous injection. An injection of chemical into a vein.

Intrinsic strength factor. Internal or built in binding power that is difficult or impossible to alter, e.g. environment for environment $Fe^{3+}$ will always attract an anion more strongly than will $Fe^{2+}$.

in vitro. Experiments performed outside the body. Laboratory bench experiments.

in vivo. Experiments on living tissue. Animal experiments.

Ionic background salt. The salt that is selected to be added to a solution of reactants and products so that together a constant ionic strength is maintained. This constant ionic strength is needed to hold fugacities constant. Nitrates or perchlorates are usually used as background salts because they minimize ion pair formation complications.

Ionic product of the medium, $K_w$. The product of the activities of the proton and hydroxide ion.

$$K_w = a_{H^+} . a_{OH^-}$$

It is constant for any given medium as long as the fugacities, temperature and pressure remain constant.

Ionic strength. $I = \frac{1}{2}\sum c_i z_i^2$ where $c_i$ is the molar concentration of an ion $i$ and $z_i$ is its charge.

Ion pairs. Between the two extremes of bonding encountered in chemistry (ionic and covalent), there exist many compounds which although they ionize (anion and cation part) when dissolved, some anions remain within the secondary hydration sphere of the cation. Such an arrangement is called ion pairing. The 'bonds' holding the ions together are a factor of $10^{-3} - 10^{-4}$ as strong as a normal covalent bond.

Ion selective electrodes. An electrode that exhibits a reproducible response to the activities of an ion in solution but is almost immune to activity changes of other ions present.

Isocitrate dehydrogenase. An enzyme that dehydrogenates isocitrates. Magnesium or manganese ions are required as activators.

Isoelectronic. Having the same number of electrons, e.g. $Mn^{2+}$ and $Fe^{3+}$.

Isotonic solution. A salt solution that has the same osmotic pressure as a body fluid.

Jahn–Teller distortion of an octahedral complex. Because of an unsymmetrical arrangement of the metal ion's electrons in an octahedral complex, not all the 6 bonds have the same length, but rather 4 short and 2 longer bonds (see Fig. G.1).

Joule (J). A unit of energy. $1\ J \equiv kg\ m^2\ s^{-2}$.

Kidneys. Two glandular organs in the abdominal cavity of mammals. They excrete urine and so remove effete nitrogenous matter from blood.

$K_w$. See 'ionic product of the medium'.

Lactalbumin. A protein found in milk. It contains all the essential amino acids.

Land-link. A means of feeding information into a distant computer by using telephone wires. The data is typed on a typewriter keyboard many miles away from the computer and the computer uses this data and any accompanying instructions. The 'answers' or output information is fed back to the user by the same communications system.

L Configuration. The mirror image of a D configuration.

Lesion. Damage, injury or flaw.

Lethal. Causes death; deadly.

Lewis acid and base. The acid is an electron pair acceptor. Similarly, Lewis defined a base as an electron pair donor.

Ligand. A complex compound is composed of a central metal ion surrounded by electron donors (anions or molecules). These surrounding donors are called ligands.

Liver. A large glandular organ in vertrebrate animals, serving to secrete bile and blood clotting reagents, to store glycogen, anti-anaemic factor and iron, to form urea and warmth, to desaturate fats, and to destruct toxins. It is usually dark reddish brown in colour.

Low spin cobalt (II) complex. See definition of high spin. Low spin cobalt's $d$ electrons are arranged [↑↓|↑↓|↑↓|↑| | ] not [↑↓|↑↓|↑|↑|↑].

Malic enzyme. Malic dehydrogenase enzyme.

Marrow. A soft vascular fatty substance usually contained in the cavities of bones.

Mass spectrometry (m.s.). A technique in which a molecule is fragmented by electron bombardment into its constituent units and these pieces are focused, by electric and magnetic fields, into different detectors depending on their weights. The distribution plot of the number of fragments having a given mass: charge ratio is called a mass spectrum.

Medicine. The science and art concerned with the cure, alleviation and prevention of disease and with the restoration and preservation of health.

Medium. In Chapter 6, medium refers to the solvent.

Menstruation. The discharging of blood from the uterus, occurring normally at intervals of a lunar month.

Metabolism. The process by which nutritive material is built up into living matter or broken down into similar substances to perform special functions.

Metal ion substitution. Many metalloenzymes can have their metals replaced by other metals that are chemically easier to identify. More details are given in Section 9.6.1.

Metalloenzyme. Enzymes that contain a metal. It is usually in the form of an ion bound in the active site of the enzyme.

Mixed complex. A metal ion complex containing two or more different ligands.

Model system. A system having the same representative structure and behaviour as the living system. An easier, or more convenient, system to study than the living system.

Moiety. A part or share.

Monodentate. A ligand that is held to a metal ion by only one donor atom is called a monodentate ligand.

Monosaccharide. A simple sugar.

Mossbauer spectra. A special type of nuclear resonance spectroscopy that involves the Doppler effect and a moving sample.

Mutation. A change that results in the production of a new species.

Myoglobin. A hemoprotein that can reversibly bind oxygen. It has a higher affinity for oxygen than hemoglobin and is used to store oxygen in muscle. It is a single polypeptide chain.

Nerves. Fibres arising from the brain, spinal cord and other ganglionic organs, capable of stimulation and serving to convey impulses.

Newborn. Just born child.

Nitrogen fixation bacteria. Bacteria in the plant and microbial worlds that reduce atmospheric nitrogen to ammonia. This ammonia can then be used to form amino acids, etc.

Nutrition. The process of providing sustenance, aliment or food.

Oestrogens. A group of female sex hormones. Some are also found in the male. In the female, oestrogens are produced in the ovary and affect the sexual cycle and some metabolic processes.

Organism. An organized body consisting of mutually connected and dependent parts constituted to share a common life.

Osmotic pressure. A concentrated solution of hydrophilic ions or molecules has an attractive force for water molecules. This force is called the osmotic pressure. The uptake of water in plants and the distribution of water on both sides of a cell membrane are under osmotic pressure controls.

Osteomalacia. Softening of bones due to the disappearance of salts.

Osteoporosis. Decalcification resulting in the softening and deformation of bones. Calcium, vitamin D or estrogen deficiencies can cause osteoporosis.

Outer sphere complexing ligand. Ligands involved in ion pairing, i.e. ligands within the secondary hydration sphere of a metal ion.

Oxidation and reduction. To increase and decrease, respectively, the oxidation number of an element. The oxidation number (a) in ionic compounds equals the electrical charge on the ion, and (b) in covalent compounds equals the charge the atom would carry if all its bonds were regarded as ionic instead of covalent.

Oxidation state = oxidation number; see definition of Oxidation.

Palliative. Superficial or temporary reliever of pain.

Parabola. The curve formed by the intersection of a cone with a plane, or, the locus of a point whose distance from a given point (the focus) is equal to its distance from a given straight line.

Parathyroid glands. These are hormone producing glands attached to the thyroid glands. They control the calcium and phosphorus balances of the body.

Partial pressure constants. For gaseous reactions, equilibrium constants are calculated upon the basis of the partial pressures of the reactants and products. Where one would use activities for solution reactions, one uses partial pressure units for gaseous reactions.

Parturition. The action of bringing forth or of being delivered of the young; childbirth.

pD. $-\log a_{D^+}$ where $a$ = activity and $D^+$ = deuterium ion.

Pepsinogen. The proenzyme of the enzyme pepsin. Stomach hydrochloric acid is the activator. Pepsin converts proteins to proteoses and peptones. Pepsinogen is a single polypeptide chain (MW 43,000) linked by 3 cystine bridges.

Periodic table. An arrangement of all the known elements upon the basis of their increasing atomic numbers and the periodicity in their electron configurations and chemical reactions.

Pernicious anaemia. Rapid or swift production of anaemia.

Perspiration. Excretion of moisture through pores of the body.

Pesticide. Insect and pest repellents and poisons.

pH. $-\log a_{H^+}$ where $a$ = activity and $H^+$ = proton.

Phase rule. A phase may be defined as any homogeneous and physically extinct part of a system which is separated from other parts of the system by definite bounding surfaces. Solid, liquid and gaseous equilibria may be considered in terms of these phases and the numbers of components involved. Such considerations are embodied in the phase rule.

Photosynthesis. The synthesis of carbohydrates from carbon

dioxide, water and the sun's energy. It occurs in chlorophyll containing plants.

Physiology. The science of the functions of living things.

Pint. 0·568 litres.

pK of an acid. Log $K$ for protonating an acid, e.g. for

$$A^- + H^+ \rightleftharpoons AH, \quad K = \frac{a_{HA}}{a_{A^-}\, a_{H^+}}, \quad pK = \log K$$

Plasma (blood). The fluid medium that supports the red corpuscles, white cells and platelets. See Table 1.2 for further details.

Platelets (blood). These are one of the several varieties of particles suspended in blood. The platelets are necessary for shed blood to clot. They are of smaller size than the red corpuscles.

Pneumonia. Inflammation of the substance of the lungs. It is called single pneumonia if one lung is affected, double if both are inflamed.

Poisoning reactions. A reaction that prevents the normal chemical processes occurring within our bodies, e.g. a foreign chemical may block the active site of an enzyme.

Polycythemia. This is the condition of having more red blood cells than usual in one's blood stream.

Polynuclear complex. A complex in which ligands are used to join 'nuclei' (actually metal ions) into a long chain. An example is given in Fig. 6.1. Polynuclear complexes can also be three dimensional.

Polysaccharide. A polymer of three or more monosaccharides, e.g. cellulose, Agar-agar.

Porphyrins. Four pyrrhole groups joined by four CH bridges. They are usually brightly coloured and form complexes with iron and magnesium.

Precursor. Something which goes before; a precedent.

Pregnancy. The condition of carrying a child; gestation.

Preservative. This has the quality of protecting a food and preventing disease in perishable goods.

Probability. Likelihood of an event occurring.

Proenzymes. An enzyme that is safe or inactive until an activator is added.

Progestin. A chemical which acts on the endometrium to induce

M

proliferation and on the uterine muscle to produce activity characteristic of pregnancy. One of the types of female sex hormone.

Prophylactic. A medicine that prevents disease.

Proton. A hydrogen atom that has lost its electron. $H^+$.

Pulmonary. Connected with the lungs.

Purgative. A chemical for cleansing the bowels, usually by means of a cathartic.

Quanta. Units or 'packets' of energy.

Rappaport units. A scale of enzyme activity.

Rash. A superficial eruption or efflorescence of the skin in red spots or patches.

Red corpuscles. Also called erythrocytes. They have no nucleus and so are correctly called corpuscles, not cells. These are the most numerous 'cells' in the blood and contain the hemoglobin. The biconcave discs have the important function of gas carriage.

Redox potential. The oxidation–reduction potential of an electrode.

Research. An investigation directed to the discovery of some fact by a careful study of the subject; a course of critical or scientific enquiry.

Ribose. A five carbon sugar that can lose oxygen to give 2-deoxy-ribose.

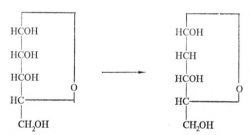

Rickets. A disease particularly incident in children, characterized by softening of bones, e.g. the spine and legs. This results in distortion and emaciation.

RNA. Ribonucleic acid. The structure is described in Section 3.5 and illustrated in Fig. 3.6. The nucleic acids are responsible for storing genetic information so that new enzymes can be built to the correct pattern of amino acids.

Screening of drugs. The sifting and elaborate testing of potential new drugs so that their effectiveness and side reactions can be established and the public protected from possible danger.

Sedative. A chemical that allays fear, soothes nerves and produces calm.

Sequestering reagents. A chemical (usually a ligand) that separates out selective ions (usually metal ions) from solution, i.e. a reagent that ties up the metal ions in such a form that any deleterious effects due to their presence are suppressed.

Serum. Watery animal fluid. The yellow liquid which separates from a clot when blood coagulates. A more detailed definition appears in the caption to Table 1.2.

Sickle cell anaemia. A form of anaemia arising from an incorrect synthesis of hemoglobin. It is a hereditary disease.

Siderosis. Accumulation of iron oxide in the lungs.

Skeleton. The bony framework of an animal body.

Solubility product. The solubility of a solid that precipitates from ions in solution may be expressed as a solubility product $K_{sp}$. For example AgCl precipitates from a solution of $Ag^+$ and $Cl^-$. Hence, $K_{sp} = a_{Ag^+} \cdot a_{Cl^-}$ where $a$ = activities.

Species. A group of animals or plants or chemicals having common and permanent characteristics which clearly distinguish it from other groups.

Spinach. A plant cultivated for culinary purposes as a vegetable.

Spleen. An abdominal organ consisting of a ductless gland of irregular form which is situated at the cardiac end of the stomach and serves to produce blood changes.

'Stat'. Held constant or static, e.g. thermostat and pH stat are instruments for maintaining the temperature and pH constant, respectively.

Statistics. The collection and arrangement of numerical facts or data. In Chapter 4 the 'statistical approach' referred to weighs up the chances of a reaction.

Steric. The arrangement in space of the atoms in a molecule.

Stimulant. A chemical that rouses or excites an organ or organism to increased activity; quickening some vital function or process.

Stoicheiometry or stoechiometry or stoichiometry. The calculation or balance of equivalent and atomic weights in a chemical reaction.

Streptococcus infections. Diseases arising from bacteria in which the cocci are arranged in chains or chaplets.

Streptomycin. An antiobiotic similar in action to penicillin. It can be used against tuberculosis and some fevers.

Subcutaneous injection. An injection of chemical into the flesh under the skin.

Substrate. A substance that undergoes a change because of the presence of an enzyme.

'Super' water. One theory of the 'structure' of liquid water is that it is a dynamic equilibrium of free water molecules, ice-like particles and super water. Super water can be obtained by vacuum distillation of ordinary water through capillaries. It has a much higher density, viscosity and boiling point than ordinary water (*see* ERLANDER, S. R., *Sci. J.* **H**, 60 (1969).

Surgery. Treatment by manual operation or instrumental appliances.

Symbiosis. Living together for mutual benefit; companionship, partnership.

Syndrome. A concurrence of several symptoms in a disease.

Synergistic drug. A medicine that cooperates with another (or others) to achieve results that neither drug could achieve singly.

Terminal groups of peptides. The primary amine and carboxylic acid groupings at opposite ends of a peptide chain.

Theobromine. The structure is described in Fig. 9.4.

Therapeutic. Medicine for the remedial treatment of a disease; healing; a curative agent.

Thermodynamics. The relationships between heat and mechanical energy and the conversion of either into the other.

Thermogram. A graph of heat liberated versus quantity of reactants added. Fig. 7.1 shows an example.

Thyroglobulin. A protein that occurs in the colloidal liquid of thyroid glands.

T.I.B.C. Total iron binding capacity. See Section 2.4.

Toxic reaction. An organism's way of showing its opposition to an introduced poison.

Toxin from cancer. A poison liberated by cancer cells. It prevents other cells from multiplying correctly. See Section 9.7.

Transferrin. Iron is transported to the liver, spleen or bone marrow as the complex of the plasma protein transferrin.

Tridentate. A ligand that is held to (a) metal ion(s) by three donor atoms is said to be tridentate.

Trigger reaction. A reaction which releases and possibly initiates another reaction.

Tuberculosis. A disease characterized by small firm rounded swellings or nodules on the surface of a body or organ.

$\tau$ values in nuclear magnetic resonance. An arbitrary unit of magnetic field strength based upon the spectra of tetramethyl-silane. High $\tau$ means high field strength.

Ultracentrifuge. A high speed centrifuge that precipitates colloidal particles in layers depending upon the sizes of the particles.

Ultra-violet spectra. The energy to promote an outer electron of an atom or molecule to another energy level is usually absorbed in the u.v. or visible wavelengths. A plot of absorption versus wavelength is known as a spectrum.

Ustilago. A parasitic fungi.

Vasoconstrictor. Produces a rise in blood pressure by constricting the muscles around the abdominal blood vessels.

Vasodepressor. Causes blood pressure to be lowered by relaxing the blood vessels.

Visible spectra. See ultra-violet spectra for description.

Vitamin $B_{12}$. The antipernicious anaemia factor. The structure is described in Chapter 8.

Vitamin D. Calciferol; the anti-rachitic vitamin. It is needed for introducing calcium and phosphorus into bones and teeth. Fish oils are good sources. Excesses of vitamin D are toxic.

Wave equation. The wave pattern of energy associated with an electron has been expressed in the form of a mathematical equation called Schrödinger's wave equation.

White cells. One type of several particles suspended in blood; they are of two types, leucocytes and lymphocytes. They are colourless and slightly larger than the red corpuscles. Their function is one of protection against bacteria, infection and the removal of damaged tissue.

Wilson's disease. The condition of a body's copper concentration control mechanism being deranged.

Zwitterion. At physiological pHs, amino acids have the ionizable proton in between the amine and the carboxyl grouping, being much closer to the former than to the latter, i.e.

$$R-\underset{\underset{NH_3^+}{|}}{\overset{\overset{H}{|}}{C}}-COO^- \text{ rather than } R-\underset{\underset{NH_2}{|}}{\overset{\overset{H}{|}}{C}}-COOH$$

Zymogen = proenzyme.

# INDEX

pD dependence of histidine complexes, 111
Pearson, R. G., 54
Penicillamine, 56, 57
Penicillin, 38, 131
Pepsinogen, 31
Peptides, 27
  bonds, 27
  chains, crosslinking in, 28
  -metal ion bonds, 51
  selective cleavage of, 134
  synthesis involving metal ions, 139
Periodic table, vi, 7
Pernicious anaemia, 17
Peroxidase, 31
Perrin, D. D., 65, 86
Perspiration, 10
Perutz, M. F., 117, 126, 140
Pesticide impurities *in vivo*, 7
pH, and enzyme reactions, 31
Pharaoh Cheops and his pyramid, 6
Phase rule, 84
Phenacetin, 38
Phenol, oxidation of, 3
Phenolase, 18
Phosphate groups of nucleic acids, 9
Phosphate(s), transfer enzymes, 13, 33, 37
Phosphatase enzymes, 56
Phosphatides, 35
Phosphoric acid salts, 10
Phosphorus
  as a ligand donor atom, 60
  era, 138
  requirements of bodies, 13
Photosynthesis, 14
Physiological effects of metal ion imbalances, 56
Pints, 4
pK determination, 73
Plasma, iron in, 16
Platelets, 4
Pneumonia, 7, 38
Poisoning
  reactions, 7
  upon a HSAB basis, 47

Pollution of environment, 55
Polycythemia, 56
Polynuclear complexes, 68
Polypeptide chains, steric considerations, 27
Polysaccharides, 34
Porphyrin(s), 59
  rings, 15, 126
  ring in chlorophyll, 12
Potassium,
  general chemistry, 9
  *in vivo*, 10
Pratt, J. M., 123, 140
Precursors, 30
Predictions of HSAB, 43
Pregnancy, 4
Preservative, 7
Principles for designing a ligand, 58
Probability, 53, 109
Processes—100,000 biological, 1
Proenzymes, 30
Progestins, 40
Prophylactics, v
Protamines, 27
Proteins, 7, 28
  donor atom's preferred metal ions, 49
  -metal ion bonds, 7, 53
  structure of, 29
  iron in, 15
Protons, 36
Pulmonary, 56
Purgatives, Epsom salts as, 13
Purines, 33
Pyrimidines, 33
Pyruvate decarboxylate, 14
Pyruvic acid, 34

Quanta, 112
Quastel, J. H., 140
Quinine, 38

Ramsay, W. N. M., 21
Rappaport units, 106
Rash, 26
Red cells, 17